CANNABIS

(SEEING THROUGH THE SMOKE)

Also by the author

Drink?
Drugs without the hot air

CANNABIS

(SEEING THROUGH THE SMOKE)

The New Science of Cannabis
+ Your Health

PROFESSOR DAVID NUTT

with co-author Brigid Moss

First published in Great Britain in 2021 by Yellow Kite
An Imprint of Hodder & Stoughton
An Hachette UK company

1

Co-author: Brigid Moss

Figures by Cali Mackrill at Malik & Mack

A CIP catalogue record for this title is available from the British Library

Hardback ISBN 978 1 529 36049 3
eBook ISBN 978 1 529 36050 9

Typeset in Sabon MT by Hewer Text UK Ltd, Edinburgh
Printed and bound in Great Britain by Clays Ltd, Elcograf S.p.A.

Hodder & Stoughton policy is to use papers that are natural, renewable and recyclable products and made from wood grown in sustainable forests. The logging and manufacturing processes are expected to conform to the environmental regulations of the country of origin.

Yellow Kite
Hodder & Stoughton Ltd
Carmelite House
50 Victoria Embankment
London EC4Y 0DZ

www.yellowkitebooks.co.uk

CONTENTS

INTRODUCTION:
WHAT'S THE TRUTH
ABOUT CANNABIS?

We are in the middle of a global cannabis revolution. For nearly 100 years, cannabis has been the most popular illicit drug in the world.

Now, every year more governments reverse the ban on cannabis – or attempt to. At the time of writing, in the US, over 200 million citizens in 36 states have access to medical marijuana and over 100 million in 17 states can legally buy recreational cannabis.[1]

Seventeen countries including Holland, Belgium and Germany have made cannabis a medicine. Israel is now offering the world's first university course in medical marijuana.

However, in the UK, the revolution has barely begun. It was only three years ago that the government finally conceded that cannabis could legally be used as a medicine, despite it being used as one for thousands of years. But because of a lack of clear direction, flawed cost arguments and ambivalence in the medical profession, only very few patients have been given an NHS prescription.

As a result, it's been estimated that over a million people in the UK are using illegal (black market) cannabis for medical purposes every day. The situation has left people lacking good information on and confused about cannabis.

You might be one of them. Perhaps you are considering taking it and want to know the real harms and benefits. Or you're already taking it, but you are worried. Or perhaps you are a recreational user or you've found out that your child is and you want to know the true nature and size of any risks.

That's why I've written this book: to tell the whole truth about cannabis, to uncover what the science says, much needed after 100 years of government propaganda against it. Most of what many of us know about cannabis is likely to come from this propaganda and possibly, more recently, from sales pitches from the new and growing cannabis industry, especially the area of CBD.

You may know me as the 'drugs czar' who was sacked by the UK government in 2009 for speaking out about the real evidence of drugs harm. I have been accused of being a drugs reformer, as if that is a bad thing. But the laws as they stand are obviously not fit for purpose. Globally, evidence shows the war on drugs has failed. The UK government has persisted with prohibition policies, which, as you can see from figures 1 and 2, have not reduced the consumption of cannabis.

The aim of prohibition is to reduce the harms caused by drugs but these policies have only ended up adding to them. This includes wasting police time, the use of public funds to prosecute people for the minor offence of cannabis possession and their life prospects being ruined. Keeping cannabis outside the law has, perversely, led to the black market selling a stronger version – skunk – and a dangerous substitute, spice.

Cannabis isn't going away, no matter how much the government tries to pretend its policy of being 'tough on drugs' is working. So wouldn't it be sensible to find a better way to exist alongside cannabis, one that causes the least amount of harm to individuals and to society, and gives the greatest amount of benefit?

As a scientist working as an advisor to the government, I didn't start out with a reform mandate. I followed one of the fundamentals of scientific research: to question everything. I felt it was my job to find the best evidence-based harm reduction policies for all drugs, including alcohol and tobacco.

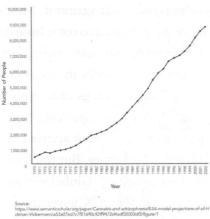

Source:
https://www.semanticscholar.org/paper/Cannabis-and-schizophrenia%3A-model-projections-of-of-H
ckman-Vickerman/ca53a07ad7c7f51690c92ff9472d4adf35003df3/figure/1

Fig 1. Number of adults reporting having ever used cannabis, 1970 to 2002

Source: https://www.statista.com/statistics/976850/cannabis-use-in-the-uk/

Fig 2. Proportion of 16 to 59 year olds reporting having ever used cannabis, 1995 to 2019

That included the bigger picture; is it logical for us to treat alcohol differently from cannabis, when alcohol causes more harm? And what government policies would lead to the greatest benefits as well as the least amount of harm?

The work I did during those years showed me that cannabis is the drug with the biggest gap between what we are told and the reality of the harm it does. Just as one example, you may be surprised by this figure, which shows the number of people who died from different drugs in the UK in 2019. You can see that cannabis is extremely low.

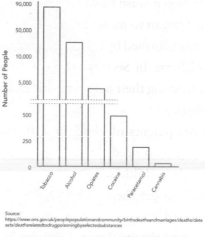

Source:
https://www.ons.gov.uk/peoplepopulationandcommunity/birthsdeathsandmarriages/deaths/data sets/deathsrelatedtodrugpoisoningbyselectedsubstances

Fig. 3 Drug related deaths in the UK

After leaving my government role, I knew I had to keep researching and talking about evidence-based policies for drugs, because it was so important. That's why I set up my charity Drug Science, and continue to campaign in this area. You can read about my experiences and my analysis of drugs policies in Chapters Three and Fourteen.

My view of the cannabis issue and the content of this book

have been informed by my time as a government advisor, and also by some of my other roles.

I trained as a doctor in the 1970s. As my fearsome senior consultant Dr Hardwick told us junior students on our first morning at Guy's Hospital, 'The purpose of medicine is to reduce suffering.' Those words have stuck with me and have underpinned my whole clinical practice.

They have become particularly poignant since I learnt about the campaign for access to medical cannabis. Despite cannabis having been a medicine for thousands of years, because of its ongoing prohibition it wasn't one I ever used as a doctor.

However, as I began to meet the people and families who'd had their lives transformed by taking it, and heard their stories – some of which are in Section Three – I realised that the government restricting their access to this medicine was – and is – a huge injustice.[2]

I have met the parents of children with severe treatment-resistant epilepsy, who need these drugs – the only effective treatment – for their children but sometimes have to pay over £2,000 a month to get them, with one having to sell their house to pay for it.[3]

That's why my charity Drug Science is pressuring the government and the medical profession to increase access. And it's why we set up Project Twenty21 – the UK's first medical cannabis data registry. At the time of writing Twenty21 has recruited over 1,000 patients, giving them access to specialist experts to ensure they get the right form of medical cannabis and at a fair price.

I also specialised in neuropsychopharmacology, which has allowed me to spend my career exploring the effects of drugs and medicines on the brain. I suspect I have given more different classes of drugs – legal and illegal – to humans than anyone else alive.

You could say that the rise of psychoactive drugs was a backdrop to my early life. I was an adolescent in the summer of love and an undergraduate in the 1970s. Psychedelics, especially LSD, were changing music, art and culture and cannabis became the hip alternative to alcohol. But I was also a child of that time in terms of the biggest ever revolution in brain science, the discovery that the brain is a chemical machine. Until the late sixties, people thought the brain was powered by electrical connections, like a phone exchange.

As an undergraduate at Cambridge, I was taught by some of the scientists who discovered how neurons work, and was party to the very first discussions on the nature of how neurons communicate with each other through chemical neurotransmission. These discoveries revolutionised not only brain science but also brain medicine. For the first time, we had a means to produce molecules that worked on the newly discovered neurotransmitters.

Drugs are just another version of this, as they also work by perturbing chemicals in the brain. I was – and am – fascinated by how much drugs change the brain. Since the first time I saw someone drunk, I have wondered, why do people do drugs? What changes do drugs produce in the brain? And what effects do these changes have on how people feel and think?

We are only at the beginning of understanding all of this. The fact that cannabis was illegal for so many years all over the world stalled this essential research. But since ten or so years ago, the pace has accelerated, and in Section Two, I share some exciting new developments in the understanding of how cannabis affects both the body and brain.

As a doctor, I also trained and worked as a psychiatrist. This, combined with my research, has given me a deep understanding of the long-standing debates around cannabis and mental health. One issue I'm sure you will want to know more about is the risk of dependence (see Chapter Twelve). The other is the linking of schizophrenia and cannabis, one of the major medical scare stories of our times (see Chapter Thirteen).

Finally, I'm writing this book as a parent. I have four children, all of whom are now grown up, but I remember the turbulence of the teenage years very well, especially as it lasted for thirteen years in our house. We all want the best for our children. We don't want them to muck up their exams, get a criminal record, or become addicted.

It's important as a parent or carer – and if you are a teacher, too – to know the truth about drugs and alcohol. Especially as the more honest information you give young people about drugs and alcohol, the more you empower them to make good choices.

In fact, I'm a great believer in empowering everyone, adults and children alike, to make choices based on good information. I hope this book gives you what you need to make up your own mind about cannabis.

HOW TO USE THIS BOOK

You may not want to read every element of the book, but to choose the parts that are relevant to you. I go into all the aspects of cannabis, what it's made of, how it works, how it can be used as a medicine and the politics of cannabis too.

To make it easier for you to find the information you want, I've broken it down into different sections.

Section One: The Cannabis Story

The history of cannabis, where it came from and how it became illegal. This section also covers my history as a government advisor and the work I did to develop evidence-based policies for cannabis.

Section Two: How Cannabis Works

This is the section to read if you want to find out what cannabis is made of, how it works and how your brain and body respond to it.

Section Three: Cannabis as a Medicine

The story of how cannabis finally became a medicine again under UK law, and why the fight is not yet finished. Plus the current evidence on the conditions cannabis works for.

Section Four: How to Minimise the Harms

We've been told a lot about the harms of cannabis. This section looks at the evidence behind those and what stands up. It also covers ways to minimise any harms, including regulation.

For cannabis- and medical-cannabis-related questions, refer to the Drug Science website: www.drugscience.org.uk.

SECTION ONE

THE CANNABIS STORY

WHY IS CANNABIS SO CONTROVERSIAL?

REALLY, CANNABIS IS just another plant. It's a relative of the common hop used in beer, and it evolved around 28 million years ago on the eastern Tibetan Plateau. Cannabis has been cultivated by human beings for 6,000 years, first for food and fibre, then later as a medicine and an intoxicant.

Looked at from the perspective of these thousands of years of humans and cannabis living happily alongside each other, the roughly 100 years of prohibition has been but a blip in time.

So why have we spent so long banning it all over the world? For most of the twentieth century, we were told that cannabis was dangerous, responsible for crime and a gateway drug to 'hard' drugs like heroin. One of the key anti-cannabis campaigners in the 1930s, Harry Anslinger, head of what was then the Federal Bureau of Narcotics in the US, said, 'Marijuana is an addictive drug which produces in its users insanity, criminality, and death.'[1] He used race and fear as weapons in his crusade against cannabis.[2]

Eighty years later the UK prime minister Gordon Brown called it 'lethal'.

The reason, of course, is THC (tetrahydrocannabinol), the chemical in cannabis that gets you high. The paradox is, it's THC that also accounts for a lot of the plant's medical effects. THC is known to reduce nausea and vomiting from chemotherapy, stimulate appetite, reduce pain and spasticity and treat anxiety and insomnia.

Cannabis also contains hundreds of other plant chemicals called cannabinoids – the one you'll have heard of is CBD (cannabidiol). At high doses, CBD can be a powerful anti-epilepsy treatment. At lower doses it can reduce anxiety and improve sleep.

As we learn more about cannabis, we're discovering that it's the combination of these cannabinoids and other active chemicals that together accounts for its powerful and wide-ranging medical effects.

When cannabis was first farmed, in 4000 BC in China, it was as a food – for its seeds and oil – as well as the plant fibre being used to make paper, rope and fabric. This original version of cannabis had low levels of THC, making it more like the modern hemp plant than the kind people smoke.

It's lost to history how people discovered its intoxicant and medical properties. It's not clear if people bred cannabis to increase its levels of THC, or if particular growing conditions led to a more psychoactive version. It's likely it was a combination of both.

Evidence of the first use of cannabis for psychoactive purposes has been discovered at a burial site dating to around 2,500 years ago. The archaeologists who examined the site think it was burned in wooden braziers in an enclosed space as part of the burial ceremony.[3]

We do have written evidence of cannabis being our oldest medicine. It comes from China and it's found in the world's first pharmacopoeia, the Pen-ts'ao Ching, which is thought to be 5,000 years old. It describes cannabis as being used for over 100 conditions, including gout and rheumatic pain. In fact, the Chinese character for anaesthesia is the same as the one for cannabis intoxication.

The spread of the cultivation of cannabis followed trade routes until it covered most of the globe, along the way becoming integrated into both religious and medical practices.

In India, it was used in medicine as a painkiller, anticonvulsant, anaesthetic, antibiotic and anti-inflammatory. And in one of the ancient Vedic scriptures, it was named as one of the five sacred plants.[4]

Evidence has also been uncovered of cannabis being used in Persian and Arabic medicine, and of its use across Africa and South America too. It had a wide range of uses, reflecting its safety as well as its versatility as a herbal medicine. You could see this as good news for the many conditions it's being tried and researched for in the present day. They included healing wounds; soothing pain including toothache, headache, earache and labour pains; and helping with epilepsy, anxiety, infection, malaria, fever and dysentery.[5]

During the Middle Ages, it's been shown that cannabis was grown in the UK. But probably due to the climate, it wasn't psychoactive. It was hemp, grown for its hard-wearing fibres, the best source of raw materials for the manufacture of maritime rope and sailcloth. In 1535, as Henry VIII was building his navy, he passed a law requiring all farmers to grow a quarter of an acre of hemp for every 60 acres they farmed.

There are descriptions of cannabis with psychoactive properties being brought back by travellers to the East around this time. Nicholas Culpeper, in his *Herbal* (1653), recommended it for earache and jaundice and also wrote that the roots, 'allayeth inflammations, easeth the pain of gout, tumours or knots of joints, pain of hips.'[6]

So how did such a useful plant turn into a pariah?

It was the increasing popularity of cannabis – as an intoxicant and a medicine – in Europe during the Victorian age that sowed the seeds of prohibition.

2

WHY DID CANNABIS GET BANNED?

IN THE 1800S, cannabis was rediscovered as a medicine in India, and brought to Britain. It became popular as an off-the-shelf remedy, in the same way as tinctures of opium and cocaine at the time. It was sold in local shops and pharmacies, and used by all levels of society.

It was championed in England by the Irish doctor William Brooke O'Shaughnessy, who became interested in it when working as assistant surgeon with the British East India Company's Bengal Medical Service.

At the time in India, cannabis was commonly taken, both for recreation and as a tonic, in three ways. There was bhang, a drink made from the leaves and stalks (the active ingredients of the plant) mixed with milk and spices – because cannabinoids are fatty molecules, they dissolve in milk better than they do in water. Slightly stronger was ganja, from the flower heads, and stronger still was charas, the dried resin. Both were smoked in pipes.

The British East India Company ran the cannabis trade in typically exploitative colonial style, profiting by selling Indian-grown cannabis back to the Indian people. This monopoly of production and sales of cannabis in India became a lucrative source of revenue that helped fund great Victorian building projects in London such as the Royal Albert Hall and also Imperial College, where I work.

O'Shaughnessy studied ancient texts and consulted with traditional doctors. He also conducted his own experiments using cannabis on animals, starting with stray dogs, moving on to cats, goats, fish, vultures and storks. He then began to treat people, reporting that cannabis was good for relieving pain and using it to treat the spasms of tetanus and rabies as well as for epileptic seizures.[1]

After he published his findings in the *BMJ (British Medical Journal)*, cannabis started to get a reputation as the new wonder drug.

Users may have even included Queen Victoria herself. Her physician, Dr John Russell Reynolds, certainly used cannabis to treat women for period and childbirth pains, publishing the definitive overview of it as a medicine in the *Lancet* in 1890. He wrote: 'When pure and administered carefully, [cannabis] is one of the most valuable medicines we possess.'[2]

As the late 1800s saw the rise of cannabis medicines, it also saw the rise of the puritan temperance movement. Agitators campaigned primarily against alcohol but wanted to ban all sources of intoxication. They saw cannabis as being used rather too much to deaden the psychological pain of miserable living conditions, rather than just to alleviate physical pain.

One direct result of temperance was the founding of the Indian Hemp Drugs Commission in 1893. This investigation was prompted by questions asked in parliament about the allegedly corrupting effects of cannabis on the mental health of the Indian people. After some months of wide-ranging and thorough scrutiny, the commission concluded that cannabis was not a cause of moral dissolution or mental problems (see Chapter Thirteen for more). And the report noted, 'As a rule these drugs do not tend to crime and violence.'[3]

Despite this, the general global trend was towards the policing and control of the use of cannabis, possibly due to the ruling classes tending to see cannabis as a source of corruption of the workers. It was banned in Egypt and South Africa in the late nineteenth century, then Jamaica, British Guyana and Trinidad in the early twentieth.

The knock-on effect of banning it at a domestic level was that international trade became illicit too. It prompted governments to try to control this trade. Another issue was that in international discussions, cannabis was often lumped in with other more dangerous drugs, namely opium and cocaine. Eventually this led to cannabis being proclaimed as addictive and dangerous as opium, under the 1925 Geneva Convention. And, three years later, to the recreational use of cannabis becoming prohibited in the UK under the Dangerous Drugs Act.[4]

However, cannabis was still officially a medicine in the UK. The beginning of the end of this came when, in 1933, the US Senate voted to rescind the law on alcohol prohibition. As a result, the 35,000-strong army of alcohol prohibition enforcement officers (now known as the Drug Enforcement Administration or DEA) faced the threat of being made redundant, along with their director Harry Anslinger.

To justify their continued employment, Anslinger created a new drug scare to replace alcohol – cannabis. His tactics included rebranding it 'marijuana' to more closely associate its use with illegal Mexican immigrants, and creating racist scare stories about its damaging impact.

Though fanciful and dishonest, Anslinger's campaign served its purpose: it created enough moral panic among the public to justify politicians banning cannabis. Anslinger's job and army were saved.

At the time, the US had major influence in the League of Nations (now the United Nations), whose 1934 report stated that cannabis had no medicinal value. To further vilify cannabis and to remove any need for its cultivation for medical use, medical cannabis was removed from the US's pharmacopeia the same year.

At first the UK held out against this outrageous denial of evidence of the value of medical cannabis, just as it did when there was a similar attempt by the US to eliminate heroin as a medical treatment. (We are still one of the few countries in the world that allow heroin as a medicine.)

The powerful anti-drug message from the US had an ally in the rapidly growing pharmaceutical industry. It was in the process of identifying the active ingredients of many plant medicines in order to create single-molecule, patentable, profitable medicines. For example, opium produced morphine that could be turned into semi-synthetic and so patentable compounds, such as codeine and heroin. But cannabis, as an unstable and very complex plant product, turned out to be much more difficult to turn into a pure medicine.

At the same time, traditional remedies and plant medicines were falling out of favour. They began to be seen as less modern, and they couldn't be injected with the newly invented hypodermic syringe. It was in the interests of this new, growing, lucrative industry to eliminate competition from herbalists and their treatments and tinctures.

It also seems likely that the newly liberated and hence rapidly growing alcohol industry would have wanted to eliminate competition from cannabis in the intoxicant market. Certainly in recent years, we have seen the drinks industry funding campaigns against US ballots designed to legalise recreational cannabis.[5]

Finally, still under pressure from the US and the UN to get in line with international policy, the UK banned cannabis as a medicine under the Misuse of Drugs Act (MDA) 1971.

The pretext was the misuse of cannabis medicines by two GPs in Ladbroke Grove in London, who were prescribing tincture of cannabis to patients and suggesting that it be dripped onto tobacco and smoked.[6]

Really, according to James Mills, author of *Cannabis Nation*, it was in part a reaction to the increasing use of cannabis by the growing Caribbean communities in British cities as well as to its association with 1960s counterculture.

So rather than just strike the GPs off the medical register, the government decided to cave in to the decades of pressure from the US and ban cannabis as a medicine.

The 1971 MDA and its amendments make little scientific sense, in particular for cannabis. The Act gave each drug a class, intended to relate to its degree of danger and so to determine the level of penalties for possession and dealing. As a (former) medicine, cannabis was also put into a schedule, which relates to safe-keeping and prescribing regulations.

Cannabis oil was put into Class A – the most dangerous – while resin and the plant leaf were put into Class B. But more perversely, considering its long history, cannabis was put into Schedule 1, the most restrictive schedule, defined as 'drugs not used medicinally'. Doctors are not allowed to prescribe substances in this schedule, and a Home Office licence is generally required for their production, possession or supply.

The upshot was that medical research on cannabis dropped dramatically, as it became expensive and difficult to get the relevant permissions (see figure 4). It cost me and other researchers thousands of pounds to obtain a Schedule 1 licence. And we had to put in a security system and keep the cannabis in a special safe bolted to the floor, a level of security above that required for very much more dangerous drugs, for example opiates.

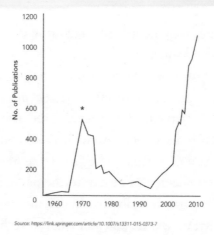

Source: https://link.springer.com/article/10.1007/s13311-015-0373-7

Fig. 4 Number of cannabis research publications since 1960

WHY DID CANNABIS GET BANNED?

Over the next 30 years, there were more than a few attempts to reclassify cannabis, both as a medicine and for recreational use. But as the propaganda of cannabis being dangerous had become so entrenched in most people's belief systems, politicians in both the major parties knew that the 'war on drugs' was a vote winner. No government could afford to stick its head above the parapet in order to support reform.

This was the situation when, in 2000, I first became involved in the arguments over the reclassification of cannabis. The following chapter describes the nine years during which I was involved in government policy, and the series of events that ended with my sacking from my government position advising on drugs.

3

CANNABIS AND THE POLITICAL MACHINE

MOST OF WHAT we've been told about cannabis being a dangerous drug is lies and myths and fearmongering.

For nine years, from 2000 to 2009, I worked as an advisor to the government on drugs, from 2008 as the government's 'drugs czar' or, more formally, chair of the Advisory Council on the Misuse of Drugs (ACMD).

I worked on three comprehensive reviews of cannabis for the ACMD, all of which included a detailed analysis of its various harms. For each of the reviews, we took evidence from a whole series of experts and stakeholders and looked at the latest scientific research. During this time, I also developed a rigorous evidence-based way to measure the harms of all drugs.

If you're not British or if you're not old enough to remember this time, the ins and outs of the political process may not strike you as relevant to whether, for example, it's safe for you to use cannabis now.

But if my experience stands for anything, it's as a first-hand account showing that what you are being told in the media or by the government may not be what the science says. It's a lived illustration of the gap between how cannabis is portrayed and how dangerous it really is.

It also shows the contempt some politicians have for evidence, demonstrating that the old Churchillian adage that science should be 'on tap, not on top' is alive and well.

At one point during 2007, the then drugs minister, Vernon Coaker, was asked by the parliamentary Science and Technology Select Committee if he understood the nature of evidence. His innocent reply? 'Of course, we seek evidence to support our policy decision.' I'm not sure he understood why us scientists fell about laughing. But what he said was peculiarly honest as well as being unscientific.

During those nine years I served on the ACMD, cannabis was a hot topic, often in the media. One reason was that in 2004, the Labour government took all forms of cannabis out of Classes A and B and put them into C, only to put them back up to B again in 2009. Another was because catching people in possession of cannabis was made into an index of successful policing. Despite cannabis being downgraded in 2004, more people were prosecuted for possession of it than ever before.[1]

Fundamentally, my experience exposed a lie at the heart of government: they don't want to tell the truth about drugs.

They want to keep drugs as a political tool, something they can use to beat their opponents and to gain votes, even if it means misleading the public.

Back when I first started working as an advisor for the government, I thought what the government said about cannabis must be based on evidence. Perhaps I was naive, but when I was invited to join the Runciman Committee to review the drugs laws in the late 1990s, I assumed the 1971 Misuse of Drugs Act was based largely on science.

The Act ranks all illicit drugs from Class A, the most dangerous, to Class C, the least dangerous. Its major aim is to provide comparative measures of drug harms for sentencing purposes.

I presumed that evidence and scientific analysis had contributed to how the law was created, at least to the level of knowledge we had in 1971. And that the ranking of different classes and schedules must therefore largely be right, if out of date. I assumed that because cannabis oil was in Class A it was very harmful, and because resin was Class B it was also harmful though somewhat less so.

The Runciman Committee was set up to review the Misuse of Drugs Act. It was funded by the Police Foundation, a charity thinktank that sponsors research into the philosophy and value of policing.

The committee chair Viscountess Ruth Runciman had been the driving force behind persuading the then prime minister Margaret Thatcher to adopt the HIV-prevention policy of needle exchanges for heroin users. This was quite a feat, considering Thatcher's staunch anti-drugs stance. The rest of the committee were impressive thinkers too, including philosopher Bernard Williams, writer and journalist Simon Jenkins and Rudi Fortson, one of the leading drugs lawyers in the UK. It was a major piece of work.

My work on the committee marked an escalation of my career-long campaign to find out and communicate the truth of the harms of drugs including cannabis.

During Runciman, I developed my first scale to measure this. The scale was divided into three sections: physical harms, addiction risk and harms to society. Each of the three sections had three questions that were used to rate each drug, and they were scored one to three. They included the risk of physical harms, both long- and short-term, the risk of dependence, withdrawal symptoms, social harm and medical costs. Each drug could score a maximum of 27 and a minimum of 9.

Using this method, we rated 14 drugs. We worked out that according to this ranking of harms, the classes should look like this:

Class A
heroin
cocaine
methadone
other opiates in pure form
amphetamines in injectable form
alcohol was at the boundary of Classes A and B.

Class B
amphetamines other than injectable
barbiturates
buprenorphine

codeine
ecstasy and ecstasy-type drugs
LSD
tobacco was at the boundary of Classes B and C.

Then **Class C** would contain:
cannabinol and cannabinol derivatives
benzodiazepines
cannabis

As you can see, cannabis came out as the least harmful, and heroin the most. We consulted the members of the Royal College of Psychiatrists, who largely agreed with our ranking.[2]

In the end, the committee's key recommendations were hardly radical as we didn't suggest any major changes to the Act. We concluded that some drugs were in the wrong classes and should be moved in line with evidence of their harms. Cannabis, we said, should move from A and B to C.

The report also said that the law on cannabis causes 'more harm than it prevents', particularly by criminalising large numbers of otherwise law-abiding young people, in particular those in minority ethnic communities.

The government seemed to agree; more than half of MPs questioned in a BBC survey said they were in favour of relaxing the laws on 'soft' drugs.[3]

And to the committee's surprise, most of the suggestions were accepted by the public and media as being sensible and evidence-based. Amazingly, a number of newspapers that were usually anti-drug reform, in particular the *Daily Mail* and *Daily Telegraph*, even criticised us for not going far enough.

We on the committee assumed policy change was bound to follow, especially when the House of Commons Select Committee called for cannabis to be moved to Class C as part of a refocusing of drugs policy towards education and harm reduction. The home secretary at the time, David Blunkett, asked the ACMD, the expert committee that advises on the 1971 Act, to formally review cannabis scheduling.

By this time I had joined the ACMD, as chair of the technical committee. But at the first meeting, I was disturbed by what I heard. The members were an experienced and talented mixture of academics and clinicians, but the process of assessing harms was opaque and unstructured.

I was surprised to hear statements that sounded more emotion- than science-based, such as 'Ecstasy will never be downgraded while I am on this committee.' I agreed to chair the group provided I could instigate a systematic review of the drug classification system using a defined and transparent process.

In 2003 the ACMD confirmed the Runciman analysis and recommended that all forms of cannabis should be Class C.[4,5]

The home secretary then announced the upcoming change in legislation. We on the ACMD assumed it was job done.

Then, just before the law was changed, for reasons unknown, some right-leaning newspapers, in particular the *Daily Mail*, turned against reclassification.

But Blunkett stuck to his guns and in January 2004, cannabis was put into Class C. While the new legislation may have looked progressive, its result was quite the opposite. To appease its critics, the new penalties for supply and/or dealing for cannabis were pushed up from 7 to 14 years. And that made them exactly the same as those in Class B, where most cannabis products had just been moved from.

The attacks on cannabis continued in the press. Blunkett resigned after allegations he'd used his position to fast-track a visa application for his former lover's nanny, and was replaced as home secretary by Charles Clarke.

There then followed the most frustrating few years, as the government repeatedly tried to get cannabis reclassified, ignoring and rejecting ACMD advice against doing this.

I can only assume Clarke and the two home secretaries who followed were being pressured by the Labour leaders, Tony Blair and then Gordon Brown, to get cannabis moved out of Class C back into B. Why they wanted this was not openly explained at the time.

With hindsight I think both prime ministers wanted to have cannabis reclassified primarily to keep the votes of the majority of the public. Most people accepted the dominant dogma of the time, that drug harms could be reduced by drug users being punished. It follows that in order for a Labour government to keep the support of those politically in the centre, they had to be seen to be as 'tough' on drugs as the Tories.

In 2005, Clarke asked the ACMD to review the status of cannabis for a second time, in case we had got it wrong on first analysis! The reason given was the claim of increasing evidence of a link between skunk and mental illness.

During the second ACMD review, we looked at new evidence, listened to more experts, then produced a report and a recommendation. Although the decision was not unanimous, the vast majority of the committee wanted cannabis to stay in Class C.

But Clarke, the government and the now mainly anti-cannabis media were not happy with our recommendations. Clarke came up with a compromise: he suggested that maybe it was the inflexibility of the 1971 Act and its three classes that was the problem. The Act was by then over 30 years old and had never been reviewed. So he asked the ACMD to go ahead and start a review. We welcomed this opportunity and started to take evidence from other countries with very different drug regulations.

Unfortunately, soon after this Clarke fell out with Blair over a failure to deport foreign criminals, and was sacked as home secretary.

Clarke was replaced by the Scottish ex-shipyard worker, John Reid. He took the view that – in his own words – there were no votes in drugs. Immediately, and without consulting the ACMD, he scrapped the review process, which has stayed scrapped ever since.

NO VOTES IN DRUGS

In fact, the government did likely believe there were votes in drugs – but in punishing people for using drugs. Previously, the police had been relatively relaxed about cannabis because 'stoners' didn't cause them nearly as much trouble as drunks. Now, as one of the criminal justice reforms, police were given performance targets, and the easiest way to achieve them was with the newly introduced cannabis street caution that the police could use to deal with people caught with cannabis for personal possession.[6]

The number of cannabis offences soared. And the policy – or its implementation – was clearly racist. Three to four times more black and ethnic minority people were prosecuted than cannabis use statistics would have predicted.[7]

This campaign led to over a million young people – mostly men – being given a criminal record and so created an underclass who found themselves faced with severely limited job prospects. It's very hard to get into the civil service, teaching or other professions with a criminal record.[8]

For many people with such convictions, crime – especially drug dealing – became their best employment option.

This punitive policing policy was utterly lacking in intellectual foundation and proportionality; how could a trace of cannabis be a justification for a criminal record? It enraged fair-minded people, including many rational police officers, who could see the targets were having very damaging effects on police–community relations. In the end, it was a major factor in precipitating the 2011 riots in London and other UK cities with significant ethnic minority populations. And it left a nasty stain on police–public relations that has not yet been erased.

CONTINUING THE REVIEW PROCESS

During this time, as chair of the Technical Committee of the ACMD, I continued to develop an evidence-based method to compare the harms of different drugs. My hope was that by conducting expert-led and rigorous research into measuring these, the ACMD could ensure drugs were in the right classes, so the punishment people received would be proportionate to those harms.

At every meeting, committee members assessed some drugs on the nine-point scale I used for Runciman, covering 20 drugs over several years. Members would present evidence and argue their opinions, then we would eventually come to an agreement. The results were published in the *Lancet* in March 2007. For this scale, the maximum score was 30. Heroin came top again, at 27. Second and third were cocaine (23) and barbiturates (21). Alcohol scored 18, cigarettes 16, cannabis 14 and MDMA 11.[9]

The results again confirmed that cannabis was in the right class – C.

In May of that year, Blair stood down as prime minister and, in June, Gordon Brown was elected as leader of the Labour Party and so also as PM. A few months later, one afternoon when I happened to be in the Home Office, my scientific secretary told me to come and watch Brown being interviewed on TV. He made the remarkable statement that the government was going to review cannabis classification yet again, saying it was because skunk was 'lethal'.

We looked at each other in amazement – why was the PM talking about cannabis without consulting the ACMD? And where did this evidence of lethality come from?

We never knew for sure why Brown took this line. My best guess is that he wanted the support of the *Daily Mail* in the run up to the general election and believed that giving them what they wanted in relation to cannabis as a start might help.

Brown's 'skunk is lethal' speech was the starting gun for this. It also led to the third ACMD review of cannabis in a decade.

THE THIRD ACMD REVIEW

There were three reasons given by the home secretary, Jacqui Smith, when she requested this review. And they were: fears about driving under the influence of cannabis; the rise in popularity of skunk; and the links made between skunk use and schizophrenia (all three are covered in detail in Section Four).

In the ACMD assessment, we brought in a number of experts on each topic to the committee. We also invited external experts to present their latest data to us. As a group we reviewed the data, particularly any new information since the previous two reviews, in detail.

We found that although cannabis intoxication clearly impairs driving performance, it does so in a different way to alcohol. Stoned people have more insight into their impairment than many drunk people do, so they drive less and, when they do, take fewer risks.

In relation to skunk, we concluded that it was not lethal, though it might be somewhat more harmful than traditional cannabis. And we reported that the association of cannabis and schizophrenia appeared to be correlative rather than causative (see Chapter Thirteen).

Taking all these issues into consideration, the ACMD and our additional experts decided, with a very decisive majority, to keep all forms of cannabis as Class C, as well as to give a whole series of recommendations on research and education.[10]

Inside the report, we published the results of a MORI poll of the public on cannabis regulations. A key question we asked was: 'What class do you think cannabis should be in?' The public's view was clear: the largest group (41 per cent) wanted it to stay in Class C. Also, somewhat to our surprise, we found that over two-thirds wanted it to be either in Class C or to carry even lesser controls; some wanted legalisation. In fact, more people (27 per cent) wanted cannabis to be legal than those who wanted the rules made stricter, i.e. for cannabis to go into Class B (13 per cent) or Class A (11 per cent)!

By the time the report was published, I had been promoted – after competitive interview – to chair of the full council.

So it fell to me to share the final ACMD report with the relevant members of government teams.

I was told that the Home Office agreed with all the recommendations – except one, the class cannabis should be in. When I asked why the answer was that, 'The public would not accept it.'

Playing my ace card, I drew attention to the appendix containing the MORI poll. I assumed the results would carry some weight with the government, as they were obtained with the research methodology political parties use to judge the public's views. I then heard these memorable words: 'That's the wrong kind of public.' I had to laugh because I fully understood what was meant. Having worked with the Home Office for nearly ten years by then, I knew their focus was on what the *Daily Mail* was saying about their policies and actions. We all knew that the 'right kind of public' was that newspaper's readership, not the representative sample of the general public contacted by MORI.

In January 2009, the government reclassified cannabis to Class B even though Jacqui Smith acknowledged the evidence was not there to do this. She said: 'My decision takes into account issues such as public perception and the needs and consequences for policing priorities. There is a compelling case for us to act now rather than risk the future health of young people. Where there is a clear and serious problem, but doubt about the potential harm that will be caused, we must err on the side of caution and protect the public. I make no apology for that. I am not prepared to wait and see.'[11]

I went to the press and said that this was the first time the government had gone against an ACMD decision about drug classification and their action was neither logical nor evidence-based.

I was unaware at the time, but this was the lead-up to my being sacked.

During a lecture I gave to the Centre for Crime and Justice Studies (CCJS) at King's College London in September 2009, I was critical of many aspects of the current policies, such as the lack of investment in harm reduction programmes and also the government's irrational, non-scientific classifications of drugs, in particular cannabis.[12]

A few months later, a paper I published based on that talk started a media storm. First, I found myself on the BBC morning news programme being asked what was wrong with the current ABC drug classification system. I replied that drugs were not in classes based on evidence of their harms. And that the Home Office's own work, published in the *Lancet* in 2007, had shown that alcohol was overall more harmful than cannabis, ecstasy or LSD.

A campaign to get me sacked was started by the ultra-puritan self-appointed Europe Against Drugs group. It incited the *Daily Mail* and *Daily Telegraph* to attack me as a dangerous revolutionary whose vision would destroy the moral fibre and health of the country's youth.

Alan Johnson, the then home secretary, asked for my resignation. I replied that I wasn't going to resign because I had been telling the truth about comparative drug harms, a truth that needed to be openly discussed because it had implications for policing and punishments.

I fought back, doing media interviews and putting my case. It wasn't until 36 hours later that Alan Johnson finally appeared in the media. And then he made a right hash – excuse the pun – of his Sky interview, losing his temper when accused of shooting the messenger rather than responding to the message.[13]

In parliament the following day, questioned by the Liberal Democrats, Johnson justified my sacking on the grounds that I had got it wrong on the harms of cannabis. He claimed that my argument that cannabis didn't cause schizophrenia was wrong despite the views of three expert cannabis reviews over nine years.

THE BIRTH OF DRUG SCIENCE

The impact of my sacking was felt across the UK science community. A petition to the prime minister to have me reinstated was set up by academic colleagues and attracted over 3,000 signatories. Half the remaining scientists on the ACMD resigned in protest. Within a month a further three had resigned, effectively demolishing it as a scientific organisation.

I am often asked if, with hindsight, I would have done

anything differently to have avoided being sacked. It's a difficult question: the process of fighting my corner after being sacked was enormously demanding and stressful. There were almost daily press interactions and a constant mass of email traffic that significantly cut into my clinical and research work, as well as my sleep.

But some good things did come out of it. It showed me there was a tremendous amount of public support for evidence-based drug laws, and that I was not alone in wanting these. For example, I gained nearly 20,000 supporters on Facebook and Twitter in just a few months.

This prompted me to set up my charity, the Independent Scientific Committee on Drugs, now called Drug Science. It was funded by Toby Jackson, a PhD in nuclear physics who had made his fortune developing one of the first algorithms for computer-controlled stock exchange trading. He was appalled by the way I had been treated. The membership of the new charity matched the best the ACMD had ever achieved, including some ex-ACMD people as well as other eminent senior UK experts in pharmacology and neuroscience.

Our remit was to 'tell the truth about drugs'. Its intention was – and is – to be an honest, impartial and evidence-based expert group free from any political influence. Ever since, we have provided the public, media and politicians – in any case, those few who will listen – with analyses on drugs and drugs policy.

These have included many, often highly cited, pieces of research, publishing a journal (*Drug Science, Policy and Law*), hosting research meetings and developing policy papers and a podcast.

Most recently, we set up the Twenty21 initiative, to help people who've been denied access to medical cannabis on the NHS afford private prescriptions, and to collect data on its effectiveness (you can read more about this in Section Three).

But our first major endeavour, which you'll read about in the next chapter, was to finish the work I'd started: to publish a seminal paper assessing drugs harms.

CANNABIS AND HARM: EVIDENCE VS POLITICS

IF YOU DRINK alcohol, you probably don't think twice about having a glass of wine or two, maybe half a bottle at the weekend.

Although we know it's bad for our health, as a society we have accepted that adults are allowed to decide whether they drink and how much they drink. Sure, there is government advice about sensible drinking. But drinking is so entrenched in our culture – boozy birthdays and barbecues, team after-work drinks, Saturday night pub crawls, wine with Sunday pub roasts, Christmas morning champagne – that most of us don't even question it.

We might feel sorry for people who become alcoholics, but we don't question most of the societal fallout from alcohol such as overloaded A&E (casualty) departments, street and domestic violence. Then there are the longer-term consequences: family breakdown, liver disease, heart disease, cancer, dementia. And, of course, the resulting cost to the taxpayer.

Compare this to cannabis. Its public image is very different: you might think of it as fringy, or for hippies, students or the criminal underclass. It's acceptable to have a couple of pints a day, but a joint? That makes you a stoner.

Cannabis has a stigma that's come from years of prohibition propaganda. This is definitely lessening, but it still exists. Think about how people talk about alcoholics compared to cannabis addicts. When a person is addicted to alcohol, there is often an element of blame on the person for succumbing to this (I'm not saying this is right, by the way). When someone is addicted to cannabis, the blame most often falls on the drug itself.

THE REAL DANGER

This would be correct and proportionate if cannabis was a more dangerous drug than alcohol. But it's not.

Towards the end of my time working for government, when I began to speak out about this truth, I fell out of favour. What I was saying – that the way our society treats intoxicants is not evidence-based – was a heresy.

But the more the politicians pushed back against the scientific evidence of drugs harms that we were producing at the ACMD, the more I wanted to provide and publicise a rigorous examination of them.

The culmination of this was the undertaking, with a team of experts, of the most detailed, transparent and objective measure of drug harms that has ever existed.

MEASURING THE HARMS

It began when I was still working for the government. I was approached by Professor Larry Phillips from the London School of Economics (LSE). He'd read my 2007 study (see Chapter Three). His email said: 'Well done David, but you could do it much better if you used MCDA. Happy to help if interested.'

I had never heard of MCDA, which stands for multi-criteria decision analysis. A type of decision conferencing, it's a powerful and relatively new method from social sciences that works for very complex situations with lots of variables. It combines data with expert knowledge, then turns this input into scores that can be compared.

Larry had already used it to help the UK government decide one of the great environmental problems of our times – how best to deal with nuclear waste. Surely it could help us make sensible decisions about drugs?

It's since been used by DEFRA to look at animal diseases, by NATO and the Royal Navy, and to make licensing decisions for medicines.

Together with Professor Colin Blakemore, the then head of the Medical Research Council (MRC) and a co-author of the 2007 *Lancet* paper, we gathered a group of 30 experts on drugs from different areas: chemistry, forensic science, pharmacology, toxicology, psychiatry, psychology, policy, education and the police.

First, we decided on the intoxicants we'd cover. We included those with well-established harms such as heroin and crack cocaine; legal drugs such as alcohol and tobacco and butane (lighter fluid); and drugs the media like to write about, such as cannabis, MDMA and LSD.

Together, we listed the thousands of possible harms that these drugs may cause, then condensed them into 16 categories.

Nine were harms to the user, seven to others and society, which we grouped together and called harms to others. The harms and some examples of them are given in the table below.

Harms to the user	
1 DRUG-SPECIFIC MORTALITY	Lethality of the drug, i.e. likelihood of dying every time you use it.
2 DRUG-RELATED MORTALITY.	The extent to which it shortens life, e.g. through road traffic accidents, lung cancers, HIV, suicide.
3 DRUG-SPECIFIC DAMAGE TO PHYSICAL HEALTH	E.g. cirrhosis, seizures, strokes, cardiomyopathy, stomach ulcers.
4 DRUG-RELATED DAMAGE TO PHYSICAL HEALTH	E.g. unwanted sexual activities and self-harm, blood-borne viruses, damage from cutting agents.
5 DEPENDENCE	The propensity or urge to continue to use despite adverse consequences.
6 DRUG-SPECIFIC IMPAIRMENT OF MENTAL FUNCTIONING	E.g. amphetamine-induced psychosis, alcohol intoxication.
7 DRUG-RELATED IMPAIRMENT OF MENTAL FUNCTIONING	E.g. mood disorders secondary to lifestyle or drug use.
8 LOSS OF TANGIBLE THINGS	E.g. income, housing, job, educational achievements, criminal record, imprisonment.
9 LOSS OF RELATIONSHIPS	E.g. with family and friends.

CANNABIS

Harms to others	
10 INJURY	The chance of injury to others. e.g., violence (including domestic violence), traffic accident, foetal harm, drug waste, secondary transmission of BBV.
11 CRIME	Volume of crime.
12 ENVIRONMENTAL DAMAGE	Environmental damage locally, E.g. toxic waste from amphetamine factories, discarded needles.
13 FAMILY ADVERSITIES	E.g. family breakdown, economic well-being, emotional well-being, future prospects of children, child neglect.
14 INTERNATIONAL DAMAGE	E.g. deforestation for cocaine production, destabilisation of countries, international crime and new markets.
15 ECONOMIC COST	Direct costs to the country (e.g. healthcare, police, prisons, social services, customs, insurance, crime) and indirect costs (e.g. loss of productivity, absenteeism).
16 COMMUNITY	Damage to social cohesion and the reputation of the community.

Then, using the latest evidence and with the input of experts, we ranked each of the 20 drugs for these 16 harms.

First, the group decided the most harmful drug for each harm. That drug was given a ranking of 100.

Then the group compared the other 19 drugs to the most harmful one, rated on a scale of 0 to 100. A drug that is half as harmful as the worst one would score 50, for example, and one a tenth as harmful 10, and so on.

All decisions were conducted openly and transparently in what is called the Delphi process. The experts worked out their own scores, then shared them with the group. When anyone disagreed, they had to say so and argue their case. Eventually, we agreed a consensus score for each drug.

Next, we weighted each of the 16 harms, according to their importance. This is subjective, so again we decided it in open discussion and came to a consensus.

For example, when considering the harms to others, we decided economic cost to others (15) was the most important. So we ranked this at 100. Then the group rated the other four harms to others against number 15. We found we cared about crime quite a lot, giving it a score of 80, but we only gave environmental damage 35.

Once the drugs were scored and the weightings done, they were plugged into the computer program, HiLife, which produced the ranking (see figure 5 below).

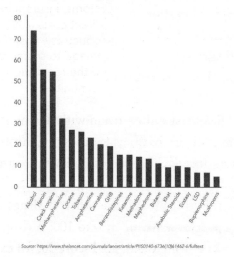

Source: https://www.thelancet.com/journals/lancet/article/PIIS0140-6736(10)61462-6/fulltext

Fig. 5. Drugs ordered by their overall harm scores

The now-famous graph (shown on the previous page) has become a meme for advocates of drug policy reform, and has been cited more than 1,600 times in research papers.[1]

The results surprised me. The 2007 study had rated alcohol as fifth most harmful out of 20; but in this study, it was the most harmful. I had long argued that reducing the harms of alcohol should be a major target for government. This was proof I was right to do so. Cannabis came eighth, evidence that I was right, in the six years leading up to the study, to argue against the reclassification of cannabis from Class C to B.[2]

The reason alcohol comes so high up the ratings is clear. The MCDA approach gave much more weighting to the impact of drugs on society, on the lives of others rather than on the user. And as we know, alcohol is by far the most harmful drug to others. We see this every day on the news, with stories about violence, drunk drivers and alcohol addiction.

The analysis also showed alcohol to be more than twice as harmful as cannabis to users, and five times as harmful as cannabis to others.

In fact, alcohol is more harmful than cannabis in most ways in which it can be measured; in particular it scored way higher on physical health, overdose risk and violence to others.

There was only one harm where alcohol didn't score higher, and that was dependence, where cannabis and alcohol were equal.

Basing drugs policies on harms

The paper was accepted and quickly published by the *Lancet*, and got quite a lot of attention.

I was interviewed about it on Radio 4's *Today* programme, where the other participant was Peter Hitchens. He said what we had done was not scientific and called me a 'ninny brain'. This caused uproar among Radio 4 listeners, many of whom complained on my behalf, and Hitchens was subsequently banned from similar programmes, though he's since been let back on.

TOBACCO VS CANNABIS

Tobacco has a very high related risk of chronic illness and mortality, from lung cancer and lung disease to heart attacks. Even when smoked, cannabis is much less harmful than tobacco for lung conditions, possibly because it burns at a lower temperature and produces fewer carcinogens. Studies looking at people who smoke tobacco and cannabis with and without tobacco suggest it's less harmful to the heart, too.[3]

Tobacco is also more addictive: up to 60 per cent of people who ever smoke become tobacco-dependent.[4]

The figure for cannabis is 8 to 10 per cent, although this may be growing due to the rise of higher-strength THC versions (see Chapter Ten).

On the other hand, cannabis does have a bigger impact on the brain than tobacco, so it scored worse than tobacco on mental function. And, again unlike tobacco, the effects of cannabis are more likely to impact your relationship with family and partner too.

The best way to know that a study is valid is to get an independent group to replicate it. And a couple of years later, the

EU's Department of Justice funded us to repeat the analysis on the same 20 drugs, this time in Brussels, with a different group of 30 experts from 20 European countries.

The new European experts adjusted the rankings of every one of the 16 criteria of harm. They also changed all the weightings, deciding that drug-related mortality was the one they cared most about.

But once the computer had done its work, the rankings were almost indistinguishable from the UK data. This tells us the MCDA drug harm assessment process is very robust, and so is meaningful in terms of directing policy. The research has been validated in two further groups, too.

Based on this analysis, it does not make sense that cannabis, being as it is less harmful than alcohol, is illegal. However, governments persist in picking one or two examples of harm – for example mental health, or crime – to justify keeping cannabis banned and not making it available as a medicine.

Now you know how cannabis compares to other drugs in its level of harms, the next section looks at the make-up of the drug itself and how it produces its effects in the body and mind.

SECTION TWO

HOW CANNABIS WORKS

5

WHAT IS IN CANNABIS?

ASKING WHAT CANNABIS contains is like asking what flavours there are in wine, it varies so much. The specific chemical make-up will depend on the strain, the growing conditions and how it's processed. And the way you take it will change its effects, too.

Besides THC and CBD, there are more than 700 chemicals in cannabis. There are over 120 that are similar in chemical structure to CBD and THC, called phytocannabinoids or just cannabinoids. Examples of these 'minor' cannabinoids are CBG, cannabigerol, and CBN, cannabinol.

The plant also contains terpenes and flavonoids, which are found in other plants too, and are responsible for aromas and colours. People usually expect that the characteristic smell of cannabis is due to the THC content, but in fact it's due to the terpenes. The main terpenes include myrcene, which is musky, and produces a sedative effect; linalool, which is spicy-floral and also relaxing; pinene which smells like Christmas trees and can be stimulating; and caryophyllene, which smells spicy and peppery and is anti-inflammatory.

The mixture of terpenes in each variety creates its taste and smell and probably also has some effect on whether you become more alert or relax more; become insular or chatty; laugh, eat, get giggly or hungry, or all of the above.

There's research going on that shows that cannabinoids and other molecules work together to produce the many and varied effects of cannabis. This is called the 'entourage' effect and it explains why, when used medicinally, whole plant extracts can be more effective than pure THC or CBD.

THE MANY FORMS OF CANNABIS, LEGAL AND ILLEGAL

FORM: LEAF, FLOWER, HERB

These are all parts of the plant, dried. The amount of CBD and THC will vary depending on the plant, as well as the part used. Hemp plants contain mostly CBD, skunk plants contain mostly THC and traditional cannabis is somewhere in between. Flower is the actual dried flowers of the plant and has the highest concentration of THC. It's often used medically for breakthrough pain, i.e. pain that breaks through traditional painkillers.

WHAT IS IN CANNABIS?

On an Amsterdam coffee shop menu, you might see strains described as sativa or indica. Sativa comes originally from Southeast Asia and Central America. It's a tall, skinny plant with a thinner leaf, said to be higher in THC and more of a stimulant. Indica originated in the Middle East and India, is shorter and bushier, and was thought in the past to be higher in CBD; so it was used as more of a sedative and pain reliever. In the 1800s it was secretly brought into South Africa by indentured workers from India, who used it to deaden the pain of their oppressive work and help them sleep.

These differences probably used to reflect significant variations in terpenes as well as cannabinoids. But now, due to lots of interbreeding, there is much less of a clear difference between strains. Cannabis plants are like dogs – different kinds can interbreed and produce new hybrids that are amalgams of both. Besides hemp, most modern cultivars are hybrids.

Medical or non-medical? Can be both.
How it's taken:
Either smoked in a spliff on its own or smoked in a spliff with tobacco (NB: this is not a good idea as tobacco leaf is more harmful to health and more addictive than cannabis leaf). It can also be vaporised, especially the flower, or smoked using a bong – a water pipe – or another kind of pipe.

FORM: OIL
Extracted from the plant using butane or supercritical (liquid) carbon dioxide. CBD extract usually comes in an oil form, and also in different strengths.

There is also a newer, super-strong version of THC, an oil that's up to 90 per cent pure that's been dubbed 'cannabis crack'. This can also come in a solid crystalline form that looks like slivers of toffee.[1]

Medical or non-medical? Can be both.
How it's taken:
The oil is either dripped under the tongue, swallowed, smoked in a pipe or put in an e-cigarette pen.

FORM: RESIN, HASHISH OR HASH
More concentrated than the leaf, hash is made by processing the dried resin produced mainly on the flowers by protective hairs called trichomes. This is then condensed into a block. Higher in THC than the plant itself.

Medical or non-medical? Usually non-medical.
How it's taken:
Smoked in a spliff with tobacco. Hot knives, pipe. Eaten in brownies.

FORM: SKUNK
The generic name for high-THC-content smokable cannabis plant. It's grown under special conditions to maximise THC production, which also drives down the amount of CBD the plant can make. The term skunk originally referred to a specific strain of very smelly cannabis but now has come to be used for all forms of cannabis with over 10 per cent THC concentration.

WHAT IS IN CANNABIS?

Medical or non-medical? Non-medical.
How it's taken:
Smoked, vaped, pipe, bong, buckets.

FORM: HEMP

This is the form of cannabis that's very low in THC. It used to be farmed only for its fibres, used in the manufacture of cloth, paper and rope. The seed oil can be used to make soap, cosmetics, supplements and as a food. Hemp is now the main source of CBD oil.

Medical or non-medical? Can be both.
How it's taken:
CBD usually comes in the form of an oil that is dropped under the tongue, but it can also come as gummies, cream or a spray.

FORM: CREAM

Studies are ongoing on CBD creams for skin conditions, as well as arthritis inflammation and joint pain. The idea is, if it goes through the skin it will work before it's broken down by the liver.[2]

Medical or non-medical? Both.
How it's taken:
Applied to the skin.

FORM: DRINK

The first cannabis drink was the Indian drink bhang; there are records of it being made as long ago as 1000 BC. Now companies are looking to cash in on the non-alcoholic and herbal drink market by making CBD-infused spirits.

Medical or non-medical? Non-medical.
How it's taken:
As a drink.

FORM: SPICE

This is the catch-all name for synthetic cannabinoids – so called because they are man-made, not plant-derived. These are nothing like cannabis, except that they also bind to the CB1 receptor in the brain (see Chapter 6). Most are illegal. Each of the many different variants has a different molecular structure and different effects, which can include extreme violent intoxication, seizures and heart attacks (see Chapter Fourteen).

Medical or non-medical?: Non-medical.
How it's taken:
Smoked after being soaked into paper or onto a herb. Because the spice molecules are not detected by tests for plant cannabis they have become the drug of choice in prison, where prisoners are routinely tested for drug use.

FORM: CANNABIS-BASED PRODUCT FOR MEDICINAL USE (CBPM)

There are four cannabis medicines in the UK, as below, three of which can be prescribed, although there are severe restrictions on this. Some doctors can also prescribe whole cannabis plant extracts, although in practice these are even more difficult for patients to access – see Section Three for more.

Nabilone is a synthetic drug similar to THC, prescribed for sickness and vomiting during chemotherapy when other treatments have not worked.

Dronabinol is a synthetic medicine identical to THC. It is not available as a licensed medicine in the UK.

Sativex is a tincture of 1:1 THC and CBD with about 3mg in each dose that is sprayed into the mouth. Licensed for people with multiple sclerosis-related muscle spasticity that has not got better with other treatments.

Epidyolex is a pure CBD medicine that's taken under the tongue. It can be prescribed for patients with two rare forms of epilepsy: Lennox–Gastaut syndrome and Dravet syndrome – and in the US for epilepsy in people with tuberous sclerosis.[3]

Medical or non-medical? Medical.

How it's taken:

Varies. Usually mouth spray or oil that's dropped under the tongue.

THE STORY OF SKUNK

Skunk was originally developed in the Netherlands by users of recreational cannabis who were looking for something stronger. A hybrid of two strains, it was bred to be super high in THC.

The name comes from the fact that it reeks like – you've guessed it – a skunk. The smell isn't due to the high THC levels, but to its particular mix of terpenes, the oils that give cannabis its taste and smell and are now thought to affect the high, too. The most prominent one in the mix is probably musky myrcene, along with pinene and caryophyllene.[4]

We now use the name skunk for any strong plant-based cannabis that's over 10 per cent THC. Traditional cannabis is usually 3 to 4 per cent. Skunk now dominates the market. In 2005, 51 per cent of cannabis seized by police was skunk, in 2008 it was 85 per cent and by 2016, it was 94 per cent.[5]

WHAT IS IN CANNABIS?

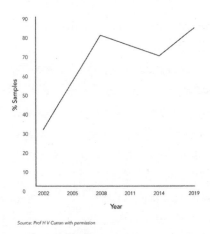

Fig. 6 The increase of high THC / no CBD cannabis strains over time

Why did it end up being close to the only product on the market? When cannabis was banned, it followed the same trajectory as alcohol during the American prohibition era: it got stronger. During prohibition, people started to drink methanol and hooch, more toxic and harder forms of alcohol.

In the UK in the early nineties, as we saw in Section One, the Labour government wanted to be seen as tough on drugs. One of their strategies was to target imports of cannabis. Customs officials made significant seizures of resin and herbal cannabis coming from traditional supply countries such as Morocco and Lebanon. At the same time, a crackdown by the Moroccan government destroyed much of the local crop.

This led to a shortage of cannabis in the UK and suppliers saw an opportunity to grow in the UK, enabled by new technology. By growing only female plants under high-strength UV lights at high temperatures and feeding them the right nutrients via hydroponic systems, growers found they could not only make the plants grow faster, but also hugely increase their THC content. And when the plant makes a lot of THC, it makes a lot less CBD.

If you were a drug dealer, you can see how skunk would make commercial sense. After investing time and money in manufacturing, the strongest product would maximise your profit. And it'd keep your consumers happy too, as they'd likely want more bang for their buck (or to get more smashed for their cash).

The new UK growing market was, at first, dominated by Vietnamese gangs. More recently, it's been taken over by British gangs as well as Dutch and Albanian ones.[6]

The business model was – and is – to take over empty properties: houses in suburbia, flats in towns, empty commercial buildings. The gangs black out the windows and fill them with plants, hydroponics and lights, usually stealing the massive amounts of electricity required to run the operation.

But the gangs' profits come at a human cost. They enslave people – often children trafficked from Vietnam – to live in the farms and look after the plants. The NSPCC have dubbed UK-produced cannabis 'blood cannabis' because while you might think of cannabis as a less harmful drug, for these children it is life-destroying.[7, 8]

As you'll read later, skunk is likely worse for the people who use it too than traditional cannabis. It increases the risk of dependence (see Chapter Twelve) and has worse effects on brain health (see Chapter Seven), due to high THC but also low CBD, as CBD may be protective. The implications of this are profound. The original policy of targeting imports of hashish, designed to reduce cannabis harms, has ended up doing the opposite.

6

WELCOME TO YOUR ENDOCANNABINOID SYSTEM

IF ALL THE hype turns out to be true, cannabis could turn out to be a genuine panacea. Probably no other drug in history has been used for so many conditions. And currently cannabis is being researched for a broad range of potential applications, from sleep issues and anxiety to seizures and pain, appetite and sociability – and even cancer.

It's only been relatively recently that we've discovered why cannabis may reach parts other drugs can't. Most drug or neurotransmitter receptors we've found are in a particular part or parts of the brain or body. For example, dopamine receptors are located in just a couple of brain areas, such as the basal ganglia – which is why in Parkinson's disease the impact is largely on movement and not vision or touch.

Cannabis is different. Its constituent chemicals fit into receptors spread all over the brain and body. They act as keys to lock into the endocannabinoid system, a bodily system that existed long before human beings started using cannabis.

The full extent of this system is still unknown, but its primary function is thought to be to bring balance to other systems of the body and brain. In fact, although you may not have heard much about it, it is as vital to health as all the eleven major organ systems, from the skeletal and muscular to the nervous, cardiovascular and digestive. Indeed, the endo-cannabinoid system plays a crucial role in ensuring the healthy functioning of each of those major organ systems.

Cannabis is a tricky plant to research in the lab, as its chemicals are large, unstable and don't dissolve easily in water, which makes them hard to work with. That's why THC and CBD weren't isolated from cannabis until the 1960s, whereas isolates from other plants such as cocaine, caffeine and morphine were all identified two centuries ago.

After the discovery of THC and CBD, there was a 20-year gap in research breakthroughs. This was due not only to cannabis being made illegal, which made research frowned upon and difficult to get funded, but also pinpointing the brain mechanisms involved turned out to be elusive.

It wasn't until 1988 that the first piece of the puzzle that would turn out to be the endocannabinoid system was discovered. This was at St Louis University Medical School by Allyn Howlett and William Devane, who first found the receptor for THC.

WELCOME TO YOUR ENDOCANNABINOID SYSTEM

The breakthrough happened like this: first, extracts of rat brains were treated with radioactive-labelled THC, then the researchers analysed the binding of this THC within the brains. The results revealed that there was a specific binding site protein (or receptor) for THC in the brain. Subsequent imaging experiments revealed the places where the THC stuck – and this, surprisingly, turned out to be all over the brain. Until that point, most other receptors had been shown to be confined to specific areas of the brain. But this newly discovered receptor, which they named CB1, was everywhere.

In brain research, once you've found a receptor, you can assume the body makes a chemical that will fit into that receptor. For example, the discovery of opioid receptors rapidly resulted in the identification of the brain's natural opioids – the endorphins.

So began a search to find the body's own cannabis-like chemical. Four years later, it was discovered by Raphael Mechoulam at the Hebrew University of Jerusalem, who's been dubbed the godfather of THC (with William Devane and Lumír Hanuš working alongside him). They called this first endocannabinoid anandamide – 'ananda' being the Sanskrit word for bliss – a nod to its connection back to THC.

More recent research has shown the name to be apt: it's thought that anandamide is (at least partially) responsible for the buzz you get from exercise, the so-called runner's high.

The discovery of a second cannabinoid receptor was a direct result of an exciting step-change in neuroscience methods in the late 1980s. That was the revolution in genetics that allowed us to read the human genetic code and so identify all the genes in our bodies. The researchers used these new tools to carve out a new route to discovery, like starting at the end of a treasure hunt and working backwards.

First, they worked out the gene that coded for the CB1 receptor. Then they hunted around the body to find other genes that were similar to this one. Once they found one, they looked at the proteins that this gene coded for – and discovered it made a second receptor protein, which they named CB2.

This is where it got really interesting: it turned out CB2 receptors are active in cells found all over the body, especially in the immune cells, on macrophages in the blood and skin and in the spleen. Researchers hadn't expected CB2 to be body-wide too. It was apparent that the endocannabinoid system was a much more important system than one designed just for getting high.

Two years later, Mechoulam and his team pinpointed a second internal cannabinoid, 2-arachidonly glycerol (2-AG). Like anandamide, this one also fits into both CB1 and CB2. But this new endocannabinoid turned out to exist in the body at levels hundreds of times higher than anandamide.

Bringing us up to the present, we now know that the two endocannabinoid neurotransmitters – anandamide and 2-AG – attach to both CB1 and CB2 receptors, at least partially.

Anandamide seems to be produced on demand by stress and to affect mood. We have less understanding of its function than we do of 2-AG, which appears to work like the oil in a car, keeping the brain running smoothly.

The importance of the endocannabinoid system is becoming clearer too. It's now known there are more cannabinoid receptors in the brain than all the receptors for dopamine, serotonin and noradrenaline put together. And that these receptors form part of a family of receptors that, in evolutionary terms, are ancient. They originated in bacteria billions of years ago, when their function was to turn light energy into body energy. Life itself has been driven by this ability.

In the interim, they have mutated so they now act more like the on–off switches of cells, keeping our system in balance. They regulate numerous processes, from pain and stress responses to memory and mood, metabolism and appetite, sleep, immune function and fertility.

ENDOCANNABINOIDS: THE BRAIN'S BALANCING ACT

What's also fascinating about endocannabinoids is that their make-up is so different to other neurotransmitters. This explains why, when the researchers first went looking for them, they couldn't see them.

A useful way of thinking about the brain is as a ginormous telephone exchange, and each of its roughly 200 billion neurons as a telephone wire. The information that travels around the neuron wires creates our thoughts, functions and processes, from digesting and thinking to remembering. It travels along the wires via electricity, but it jumps the gap – the synapse – between using chemicals called neurotransmitters.

You'll already know some of these – serotonin and dopamine, for example. The major neurotransmitter players in the brain are glutamate, which acts like the on switch for the brain and wakes us up, and GABA, which does the opposite. These are made and stored in the neurons, ready for use.

When researchers looked for a store of endocannabinoids, they couldn't find one. But then they worked out that endo-cannabinoids aren't stored at all. Instead, they are made and released on demand.

Thinking again about a telephone exchange, each of the electrical wires will be insulated, in order that they don't short-circuit each other. In your brain too, each neuron is insulated by a fatty membrane. When a neurotransmitter – usually glutamate – jumps from one neuron's synapse to another, the membranes get shaken up. Some of the fatty content is released, and that becomes the endocannabinoid.

Endocannabinoids do something else that's unlike other neurotransmitters: they travel back across the gap to the original synapse. When it arrives, it gives a message to stabilise or blunt any further release of glutamate.

So, rather than simply raising or lowering the level of gluta-mate, endocannabinoids do both, depending on your existing level. This is how your system works to stay at the optimal level for your brain functioning.

Almost all brain activity results in the production and release of endocannabinoids. And they can be made by almost any cell in the brain, which is why the receptors are all over the brain too. As well being made by neurons, they can be made by astrocytes, a kind of caretaker immune cell that provides a nurturing environment for neurons, crucial to the smooth running of all brain functions.

You need the right quantity of endocannabinoids in the same way as you need the right quantity of vitamins. Both too little and too much can be a problem. It's thought some medical conditions may be caused by the endocannabinoid system being out of balance, either underactive or overactive. Let's take epilepsy as an example. We know that when the endocannabinoid system is underactive, it's not doing its necessary balancing act to lower levels of glutamate. And we also know that too much glutamate in the brain causes seizures. This might explain why medical cannabis works to control some types of epilepsy.

Cannabis itself also works as an adaptogen, with various ways of bringing the endocannabinoid system into balance. CBD appears to moderate and dampen down excessive activity on both cannabinoid receptors. THC may also have a similar rebalancing effect.

THE STORY OF THE ANTI-CANNABIS DRUG

While you can feel the effects of, say, smoking a joint or eating a brownie, it's not clear if we feel the effect of endocannabinoids being released. You've likely felt the runner's high, but that involves other brain chemicals too.

One reason is that it's technically difficult to measure the release in the brain of endocannabinoids. Currently, in humans, the best we can do is to measure them in spinal fluid. But this involves a spinal tap, not the most pleasant experience for volunteers or patients, so it's rarely done in studies.

There is one class of drugs that suggest the feelings endocannabinoids might be producing. These are the antagonists, which bind to and fully block the receptors.

You'll have heard of other drugs that work like this: beta blockers, for example, block noradrenaline receptors and so stop part of the physical stress response. Antagonists also bring on withdrawal. One well-known example of this is naloxone, the antagonist to morphine and other opioid medicines. By blocking the receptors, it also blocks the effects of the body's natural opioids, the endorphins, too.

The same is true of cannabis antagonists. They block the effects of THC in cannabis, as well as your body's own endocannabinoids.

The first of these antagonists was developed in 1994 by a French company called Sanofi (now Sanofi-Aventis), who named it rimonabant.

Surprisingly, when given to healthy people there were no obvious or immediate effects or problems.

This makes it different from giving the antagonist for other neurotransmitters. Give the antagonist for the 'on' chemical glutamate? You fall asleep. Give the antagonist for the 'off' chemical GABA? You get anxious or have a seizure.

The effect of rimonabant was a lot more gentle: over time, people seemed to have a general lowering of mood. At first, rimonabant was developed as a possible treatment for schizophrenia. This was based on the theory that THC makes people have psychotic symptoms, so a drug that blocks the THC receptor might do the reverse. This turned out not to be true.

Sanofi looked for other uses. Their thinking was: THC causes the munchies, so might blocking the THC receptor help reduce appetite? So rimonabant was tested for weight loss. It seemed to work in some patients, and it was licensed by the European Commission in 2006 for weight loss in people with a body mass index of over 30, or over 27 for those with other health complications.[1]

Then, after reports of depression and suicidal thoughts, and some suicides, the European Medicines Agency recommended it no longer be prescribed to people with any history of depression, and the US Food and Drug Administration (FDA) refused to license it. It could be the decision was somewhat political; at the time there was a general antipathy to France after the French special services blew up the Greenpeace ship the *Rainbow Warrior*.

After the FDA decision, Sanofi decided to withdraw rimonabant completely, worldwide. Then, sadly, Merck and Pfizer, who had similar antagonist drugs, withdrew theirs too, even though their data didn't show similar negative mood effects. They were likely worried that extra scrutiny would scare doctors off using their compounds and so make them unprofitable. And that they might be liable to litigation from any patients who did experience any mood changes.

But whatever the motivations, we lost a whole class of potentially useful drugs. They could possibly help treat cannabis dependence by blocking the 'high' of cannabis in the same way naltrexone does for alcohol and heroin. They could stop psychotic symptoms after taking cannabis. And they could even save lives by treating the psychosis, violence, seizures and heart attacks caused by spice (synthetic cannabinoids), as explained in Chapter Fourteen.

HEADING INTO THE FUTURE

There isn't a lot of incentive for pharmaceutical companies to investigate if, or how, medical cannabis extracts work. That's because cannabis doesn't fit the usual drug development model: making or extracting a single compound that can be easily trialled and patented.

So instead of investigating cannabis itself, drugs companies have taken a second route; they are working on developing drugs that increase and reduce the body's levels of its own endocannabinoids.

One way to do this is to manipulate the enzymes that make or break down endocannabinoids. Pharmaceutical companies like working on enzymes because they are what's called very 'druggable'. That is, they make very good drug targets because it's relatively easy to make small molecules – aka drugs – that can block them.

How might this work? If you think of the body's endocannabinoid level as being like the level of your bathwater, blocking the enzymes that break down endocannabinoids is like putting in the plug.

Trials of drugs that block these enzymes are being researched for Tourette's syndrome, anxiety disorders and pain disorders.

LOOKING INTO THE UNKNOWN

There's so much we still don't know about the endocannabinoid system, which makes this a fascinating time. We do know there's more to it than we previously realised: there are more endocannabinoids, more receptors and more plant chemicals – not just from cannabis – that stimulate these receptors too.

Over the next few decades we are bound to discover a great deal more about the role of the endocannabinoid system, in health and disease. This will allow us to develop more new medicines, and better ways to use cannabis as a medicine too.

7

WHAT HAPPENS WHEN YOU TAKE CANNABIS?

I WAS A medical student at Cambridge when I tried cannabis. It was 1971 and a few of us had gone for the weekend to a country cottage in King's Lynn, Norfolk. This was around the time the Beatles had admitted to smoking a joint in Buckingham Palace, and 'dope', as it was called then, was 'hip', a key ingredient in the new music and film counterculture. There had been a boom in people taking it, although the numbers were still low. One study shows around half a million people – including me – used cannabis in the UK that year.[1]

Cannabis had been made illegal internationally in 1961 by the first UN Convention on narcotic drugs but it was still a medicine. It was part of the vibe of the time, of chilling out and reflecting on the meaning of life. That summer, it felt as if it was something a student should do, a rebellion against alcohol and the establishment. Smoking it made you feel as if you were bucking the system – if only a little. We assumed it was going to be made legal, because it seemed relatively harmless.

So there we were, in the depths of the countryside, for a weekend break from study and student protest. We'd heard of one medical student who was caught by the police with cannabis. He was told by the General Medical Council that every year after that, he would have to be assessed to make sure he wasn't an addict; and every time he changed job, he had to declare his offence. We certainly didn't want anything similar to happen to us.

I had never smoked tobacco – I had a slight wheeze, so I struggled to inhale. (As we all later found out, I had that in common with Bill Clinton.)

Almost straight away, I felt lightheaded and a bit dizzy; then everything started to seem funnier. But a friend, who we'll call S, surprised us with his reaction. He began to panic, and got extremely frightened and a bit paranoid.

I had seen peculiar reactions to alcohol before, but this reaction to cannabis was something I hadn't experienced. We drove him to hospital.

By the time we got there, S had calmed down. The doctor who saw him said he had had a panic attack, and gave him Valium. What a sight we must have been, medical students panicking about a panic attack!

As you can imagine, S never touched cannabis again. His experience didn't have any other long-term effects; he became a very senior doctor. But it was baffling to me why he had that experience and I had the opposite one. S didn't seem to be a particularly anxious person; rather he appeared extrovert and outgoing.

WHAT HAPPENS WHEN YOU TAKE CANNABIS?

How you feel when you take cannabis is not guaranteed. It's not like ibuprofen; you take that, inflammation reduces. Or even like Valium; you take that, your muscles relax. Everyone's brain has its own reaction to cannabis. For most people, it will mostly be positive, or it would never have become popular at all. For others, like S, it's pretty awful, and those people never take it again.

Why are the effects unpredictable from person to person? We don't know. We do know that as cannabis is a plant rather than a chemical, it's not an exact science. There are hundreds of different strains, all with different proportions of active chemicals. Even on a menu in an Amsterdam coffee shop, you might be given a choice of 20 or 30 varieties.

In fact, this is one of the challenges of using cannabis as a medicine. Companies that grow medical cannabis do all they can to standardise their product to create a reliable dose, by cultivating the same strain and using the same growing and harvesting conditions each time.

There are other elements that add to or take away from your buzz, if that's what you're after, as well as the other effects. An obvious one is the way you take cannabis: whether you smoke, vape or use oil drops, for example. But there is also your environment: where and when and with whom you take it. Finally, there's arguably the most crucial variable: why you're taking cannabis. If you are taking it to stop pain, for example, your experience will be hugely different from smoking a spliff with friends at a party.

71

INHALATION

Smoke a spliff and around 20 seconds later, you'll be giggling or stoned or paranoid. That's because inhalation is the fastest and most effective way of getting cannabis to your brain, due to the enormous surface area of the lungs, which allows the cannabinoids to get across the lung membrane and into your blood quickly. Five seconds later they've travelled from the lungs to the left side of the heart, and five seconds after that, they're entering the brain and hitting the brain receptors. THC goes to CB1 receptors, mainly in the brain, and CBD goes to the brain where it moderates CB1 receptor effects but also to other body sites that are not yet well understood.

Blood levels of THC peak at 15 minutes after smoking, then slowly decline until, four hours later, you're back at zero.[2, 3]

There are other ways to inhale. Using a vaporiser gives you a greater and even faster hit than smoking a joint. It heats the plant but without burning it. It's likely this may turn out to be safer than smoking, although we can't yet say for certain. There are also vape pens – not to be confused with vaporisers – where you vape CBD and THC in the same way as you do nicotine, in a propylene glycol solution. This is likely to be somewhat safer than smoking too. NB: the EVALI (Vitamin E-associated lung injury) crisis in the US was down to the solvent being toxic when inhaled, rather than the THC content – see Chapter Thirteen for details.

A hookah or bong bubbles the smoke through water, which removes some of the toxins.

As you'd expect, the bigger the dose, the bigger the effect. A group of researchers, including me, worked out a functional unit of cannabis, which we called a Standard Joint Unit (SJU) and is equivalent to what people generally take per spliff. We did this by asking people to donate joints and, surprisingly, 315 people did. We then analysed the THC content; we found that an average spliff contained 7mg of THC.[4]

UNDER THE TONGUE

As cannabinoids are fatty molecules, they dissolve easily in oil. As a medicine, cannabis oil has become more common, and CBD in particular almost always comes in the form of oil. To take it, you drop it under your tongue. The cannabis medication Sativex is a spray, which you also take under the tongue.

The oral route takes a little longer to work than inhaling: it has to go through the mouth membranes into the blood, travel to the heart, then the lungs, and only then does it get to the brain, which takes 10 to 15 minutes.

THE EVALI SCARE

This was a health scare in the US that was blamed on cannabis and e-cigarettes, although in fact the real culprit was poor science and use of unsafe vaping liquids.

It began in Wisconsin in July 2019, where cannabis was at that time illegal. Teenagers started to present to hospitals with lung problems. When the doctors treating them questioned them, quite a lot of them did not admit to taking THC, but did say they were vaping nicotine or liquids flavoured with menthol or even bubble gum.

The symptoms were rapid and laboured breathing, coughing, fever, fatigue, severe pneumonia and respiratory failure. In just over a year, there were 2,500 cases and 50 deaths.[24]

This outbreak fed into the hysteria around nicotine vaping and was used by anti-vapers to justify vaping being banned. And, when it was discovered that in a lot of cases people had been vaping THC, the illness was also blamed on this.

Finally, the US Centers for Disease Control and Prevention worked out that it was the solvent – vitamin E acetate – that was to blame, not the act of vaping or the THC.

As the products weren't licensed, and there are so many counterfeit versions, there was no control over the solvents used. We can only speculate that the usual nicotine solvent wasn't being used because THC doesn't dissolve well in it.

It's also possible that other vaping solvents could cause damage to the lungs. This is why it's really important there is a properly regulated market for all vaping solvents, whether that's for vaping nicotine, THC or just flavours.

EDIBLES

When cannabis is made into an 'edible' such as a chocolate or brownie, it takes even more time to work. And while the peak of the effect is much flatter, it lasts longer, up to six hours.[5]

After swallowing, the edible starts dissolving in your stomach acid. Some of the cannabis molecules will be broken down by this. Some cross the stomach wall into a blood system called the portal venous system, where all food products go. In there, they travel into the liver, where more of the molecules are broken down. What's left goes into the normal venous system, then to the right side of the heart, into the lungs, into the left side of the heart, then into the brain. All of this can take 30 minutes to an hour, longer if you've eaten recently, as this slows the stomach process down.

With the commercialisation of non-medical cannabis, there's a danger of edibles often coming in strengths higher than you'd expect. Whereas you can stop inhaling, once you've swallowed an edible you can't stop the effect – unless you vomit.

They also come as products you'd normally eat more than one of, like gummies. For example, in some US states and Canada, gummy bears can contain 10mg per sweet. For inexperienced users, 10mg is a big dose, when you think that a whole spliff is usually 7mg.

If you were eating normal sweets, you'd probably usually eat a handful. But if you ate, for example, five cannabis gummies, you'd be out of your head.

There's also the issue that edibles such as gummies are more likely to be taken by children, either by accident because they look like sweets or because they're easy to take. In October 2020, 14 London schoolgirls were taken to hospital after taking THC-laced gummy bears, then reporting nausea, vomiting and dizziness.[6]

WHAT DOES TAKING CANNABIS FEEL LIKE?

Most of what you notice when you take cannabis is from THC as CBD has very few noticeable effects.

The first thing THC does is dilate your blood vessels. It's this that causes the telltale red eyes of someone who's stoned. At the same time, it increases blood flow to the brain. This head rush may be part of cannabis's pleasurable effect: we know that some people like the feeling of changes in blood flow to the brain. In a more extreme form, it's what amyl nitrites, also known as poppers, do, and it's the effect behind auto-erotic asphyxiation too.

In what seems like an opposite effect, cannabis also increases blood pressure and heart rate. This might make you feel as if your heart is fluttering, racing or pounding. We don't know the physical driver for this but we do know from post-mortems that the biggest risk of cannabis causing death is to those who have pre-existing heart problems, specifically ischaemic heart disease or irregular heart rhythms. It seems that increase in blood pressure and heart rate can put extra strain on the heart in the same way that sex and exercise might, and so can lead to a heart attack.

However, this is very rare. In 2019, there were seven deaths from cannabis where no other drugs were involved.[7]

Additionally, smoking cannabis can provoke asthma.

There are no other immediate dangers to health from taking cannabis – there is more on possible long-term harms in Section Four. Because it's been illegal, unlike alcohol, we haven't been able to do epidemiological studies on the long-term effects of people who take it. But it's been used for thousands of years and is now classed as a medicine, which tells you it's been through a great deal more safety checks than alcohol. If alcohol was made for the first time now, it'd probably fail to be allowed as a food, let alone a medicine, as it's far too dangerous.[8]

What we do know is that, unlike alcohol, cannabis is almost never fatal when you take an overdose. If you take too much, too fast, or if you're not used to the effects, or if you're just unlucky, you could have what's colloquially called a whitey. This is when your blood pressure plummets, you feel sick, go pale and might even vomit. That said, you might have a panic attack and *think* you're dying, as my friend did, especially if these physical symptoms coincide with paranoia, explained below.

You're more likely to have a whitey if you mix marijuana with alcohol. If that happens to you, lie down. NB: If someone you're with is vomiting due to cannabis and/or alcohol or any other drug, don't leave them on their own in case they choke. Put them in the recovery position and consider calling an ambulance.

A small proportion of very heavy cannabis users get what is called cannabinoid hyperemesis syndrome, with the main symptom being that they keep vomiting. It's alleviated by having hot showers and baths, and stops when they stop taking cannabis.[9]

BRAIN IMAGING OF CANNABIS

As part of the Channel 4 programme *Drugs Live: Cannabis*, my colleague Professor Val Curran and I did a brain imaging scan study. It compared the effects on the brain of high-THC cannabis – skunk – with cannabis containing a mixture of CBD and THC. The level of THC in the two types of cannabis was exactly the same. We hoped this would help us understand why skunk might be causing more problems than traditional cannabis.

Because cannabis is illegal, there were plenty of regulatory hurdles. We ordered the two forms of cannabis from Bedrocan in the Netherlands. They make medical cannabis, so we could be confident of the exact concentrations.

We administered the cannabis using a type of vaporiser called the Volcano. This has a heating element that vaporises the cannabis molecules into a large bag. When it's full, the subject inhales the bag in the same way as people inhale balloons of nitrous oxide. It gives them quite a high dose of cannabis quickly, which maximises the changes we see in the brain.

There were 16 people in the study, and it was double-blind and placebo-controlled. On three different days, the volunteers were given a placebo that smelled and tasted the same as cannabis, the high-THC cannabis and the THC/CBD mixture. The dosage was around a third of a spliff. After taking the mixture, we asked the subjects to do some tasks and looked at their brains in an MRI scanner.

One of the presenters, Jon Snow, wanted to try it out. We warned him that if he hadn't been using cannabis recently, it might be a strong dose. But he insisted. He did complete all the tasks, but couldn't bear to be in the scanner. Afterwards, he said it was the most terrible experience of his life. Taking a huge hit of THC had produced a strong feeling of anxiety and dissociation that he found very disturbing. A witty headline in the press afterwards read: 'Just say no, Snow'.

The key finding of the study was that there are clearly different brain signatures produced by THC alone, and THC plus CBD. The mix produces less disruption in brain function than THC alone.

These differences help make sense of the growing understanding that CBD is protective against the worst impacts of THC, in particular how it can reduce the risk of becoming dependent.

From this, Professor Curran managed to get funding to test CBD in people who were THC-dependent, and she found that it did help them reduce their use (see Chapter Twelve for more).[10]

THE BRAIN EFFECTS

The brain usually works in a very rigid, repetitive, habitual way. Brain imaging shows that THC alters this. It releases the brain from what it usually does, disrupting the normal thinking process.

For example, musicians are trained to read a score and then replicate it perfectly. But under the influence of cannabis, this is hard to do. It seems likely that jazz came from performers using cannabis, which made them break out from normal musical notation and develop syncopation.[11]

These brain changes can give people useful insights into themselves and their relationships and bigger philosophical issues. Or they might just get the giggles, as cannabis breaks down their normal state of anxiety, helping them feel relaxed or lose their worries or raise their low mood. Being stoned heightens senses too, so music, food and visuals might feel more intense.

Brain imaging studies also show that cannabis is the only drug used recreationally that increases brain metabolism and so the uptake of glucose in the brain.[12] All other drugs, including psychedelics, dampen down brain activity.

Our most recent imaging studies have shown that cannabis disrupts key circuits in the brain, particularly the circuits that control attention and the significance of external stimuli (called the salience circuit). It also disrupts the default mode network, which gives you your sense of self. If the brain was an orchestra, the default mode circuit would be like the conductor, in charge.[13]

What might be happening with brain chemicals here? It's likely that cannabis works via some effects on serotonin. It may also have an effect on noradrenaline, which is why you become relaxed. But its main effect is probably on glutamate, the brain's on switch, which makes up around 85 per cent of the neurons in the brain.

There are possible mental downsides to using cannabis. Some people feel paranoid, confused or upset, or have strange thoughts, get anxious or even have a panic attack.

WHAT HAPPENS WHEN YOU TAKE CANNABIS?

There is some evidence that paranoia may come from cannabis releasing dopamine in some parts of the brain, which leads to overactivity of the salience circuit. THC can also produce strange feelings that you're not yourself as well as distortions of your body, visual hallucinations and imaginary scenarios that feel real. For this reason, some researchers use THC as a model of psychosis in healthy volunteers.

The higher the amount of THC, the more likely you are to have a bad reaction, and the more extreme it might be. Very rarely, the paranoia can keep going even after the person has stopped being stoned, which is called a psychotic episode, and this may require professional help (see Chapter Thirteen for more on psychosis).

It's well established with most drugs – stimulants, alcohol and opiates, for example – that how fast a drug gets into your brain has a profound effect on your subjective experience of it. The speed can sometimes be way more important than the route. It's why smoking crack gives a faster, bigger high and is more addictive than snorting cocaine. Or why a shot of vodka makes you feel more drunk than the same amount of alcohol ingested more slowly as beer. This effect is less well established with cannabis but the Jon Snow experience we described (see page 77) suggests it is real.

Ninety minutes after inhaling, you may get the munchies. We don't know the exact mechanism here. It could be due to the brain being more active and so using more glucose; we do know that if you starve the brain of glucose, you become very hungry.

In a study carried out on rats, cannabis provoked a surge in the appetite hormone ghrelin and changed the body's response to it. Ghrelin tells the body you're hungry, even when you're not.[14]

Another theory says cannabis might work directly on the hypothalamic/eating centres of the brain. And a third says that, if you're more relaxed, you may stop caring about controlling your food intake.

Paradoxically, people who smoke a lot of cannabis recreationally don't seem to put on weight. We don't know why exactly, but this may turn out to be of great interest to diabetes and obesity research. It could be because it acts as an adaptogen: that is, while it stimulates your appetite, it also resets it or satiates it. Perhaps if your mood is better, you feel less of a need to eat for emotional reasons.

Cannabis is known to promote sleep, although there hasn't been enough work done in this area to tell us why.[15]

We know more about how other drugs induce sleep: for example, that alcohol and benzodiazepines (which include Valium) change the sleep profile by dampening down the brain. But cannabis seems have a more subtle effect on sleep, possibly by disrupting the negative effects that impede sleep.

Depending on whether you inhale or eat it, the feeling of being high from cannabis will last up to four hours. As it's broken down and metabolised, some of the products are excreted and some will stay in the blood.

The next morning, you'll wake up with no headache or hangover (if you haven't also been drinking). Because cannabis doesn't have such a big effect on your body, compared with alcohol and cocaine, and because it clears from the brain relatively slowly, you don't go into a state of acute withdrawal as with, for example, alcohol or heroin.

In fact, you're likely still a tiny bit affected even when the intoxicating or high effects have worn off. Studies on airline pilots using a big dose of 20mg THC showed that not only are the metabolites detectable in blood plasma, but the (dys)coordination effects lasted 24 hours after taking cannabis (more on this in the chapter on cannabis and driving). A review of studies showed that flying performance was back at baseline by 48 hours later.[16]

Some cannabis is also filed in the body's fat stores. From there, it will leak out from the cells gradually, more quickly if you lose weight.

SECTION THREE

CANNABIS AS A MEDICINE

HOW MEDICAL CANNABIS
BECAME LEGAL

THE UK IS now one of the most backward countries in the world when it comes to medical cannabis. This is despite it being one of the last countries to ban cannabis as a medicine, in 1971. At the time, the world's prohibitionist dogma said that banning cannabis as a medicine would also stop recreational use. History shows this did not happen.

Fifty years later, while the UK stalls, the rest of the world is opening up to cannabis as a medicine. At time of writing, there are 37 US states and 47 countries that allow medical cannabis.[1,2]

In November 2018, after years of pressure, patients forced a law change in the UK, allowing cannabis to be prescribed once more. But two years after cannabis was made legal as a medicine, there were still only a handful of prescriptions made on the NHS.[3]

I've met countless patients and parents of patients who are desperate for a prescription – or at least to be able to try cannabis as a medicine.

This has left some patients suffering unnecessarily. It's left others using cannabis illegally. A survey by Dr Daniel Couch of the Centre for Medicinal Cannabis (CMC) showed that 1.4 million people in the UK are using black market cannabis to treat a medical condition.[4]

In this situation, preventing people from accessing medical cannabis is perverse. This has become my main focus: to help people get these medicines.

After all, we know cannabis has a good safety profile. Even where it doesn't treat the condition itself, evidence from the millions of people around the world who use it shows that it can improve quality of life, including in the areas of pain, sleep and feelings of well-being.

But why is change happening so slowly in this country? And what can we do to speed it up, for the benefit of people who need this medicine?

There are multiple barriers. But the main drivers are an unscientific bias against cannabis from 50 years of propaganda, and doctors not being educated in cannabis medicine. Also, the system of medicine licensing isn't flexible enough to license a whole plant medicine, there is a chronic lack of research and there's no clarity around funding for NHS prescriptions.

The situation is complicated by the fact that there are two levels of cannabis medicines. First, there are the two pure extracts Epidyolex and Sativex, licensed medicines which we have looked at.

Secondly, there is the whole plant extract, the traditional medicine. Some patients say they do better on this, possibly due to the entourage effect from all the plant's many chemicals. It's flexible, as we've seen in Chapter Five, as it comes in a whole range of forms and it allows you to adjust levels of CBD and THC to suit your condition and symptoms.

HOW THE LAW CHANGED ON MEDICAL CANNABIS

Alfie Dingley started having seizures at the age of eight months. His mum, Hannah Deacon, says that from then on his life was punctuated by up to 150 excruciatingly painful seizures a week, and regular hospital stays. Aged five, Alfie was diagnosed with a very rare form of treatment-resistant epilepsy called PCDH19.

'Watching a child you love having a seizure is the most frightening thing you can imagine,' Hannah has said. 'They go blue, they stop breathing. You're thinking: are they going to take a breath? Is this going to be it?'[5]

Alfie's doctors and parents gave him every possible therapy, trying nearly 20 combinations of drugs, as well as the ketogenic diet and immune therapy. The only treatment that helped stop his seizures was high doses of intravenous steroids, but this was not sustainable. Indeed, over time it could have killed him.

'After this treatment, he'd have three or four days of horrendous behaviour: screaming, crying, hitting, running in the road. We couldn't leave the house,' Hannah says. In 2016, Alfie was in hospital 48 times. 'He had no quality of life, and, as a family, we were desperate.'

Evidence was growing that cannabis could help children like Alfie. In the US, a CNN documentary publicised the story of a girl called Charlotte Figi. It told how cannabis helped her go from having hundreds of seizures a day to almost none. The strain that she was treated with was grown in Colorado, where cannabis is legal. It was originally called Hippie's Disappointment, because of its high CBD and low THC content, but was renamed Charlotte's Web when it became clear it was benefiting her. After the documentary aired, demand for Charlotte's Web rocketed.[6]

(Sadly, in 2020 Charlotte died aged 13; it's thought from complications related to Covid-19.)[7]

Alfie's parents asked to join a trial for the CBD-based medicine Epidyolex, but Alfie didn't fit the strict profile for the trial and so was refused.

Hannah was told there was no chance of Alfie ever being prescribed medical cannabis in the UK. 'Even after Alfie was admitted to hospital in the UK for the 48th time, I was told he would "never" get an NHS prescription for potentially the only treatment that could save his life,' she wrote in the *BMJ*.[8]

Alfie's parents decided their only option was to go to the Netherlands for Alfie to try cannabis medicines, and started raising funds. Since the decriminalisation of cannabis in the Netherlands in 1976, it has probably become best known for its coffee shops, but it's also a producer of medical-grade cannabis. That means the active ingredients of the plant are standardised and the plants are monitored for stability, humidity, microbiology, pesticides, heavy metals and toxins.[9]

So the family moved to the Hague. And after five months of being treated with cannabis oil, Alfie had his longest spell in recent years without a seizure – 41 days. Finally, Hannah says, Alfie had a quality of life. 'He still had to take anti-epileptics alongside cannabis. But whereas before he was taking five different ones and still having hundreds of seizures a week, [now] he wasn't having any. And he was able to cut down on the anti-epileptics, which reduced the side effects – aggression, behaviour problems, bowel and skin problems.[10, 11]

But when money ran out, the family were forced to return to the UK. And as Alfie's medication was illegal in the UK, they weren't allowed to bring it back. Back at home, Alfie's seizures returned with a vengeance.

'When I came back to the UK in February 2018, I knew I needed Alfie to have an NHS prescription and I needed guidance through the strategy. So I linked up with End Our Pain, a group that had been lobbying on medicinal cannabis and multiple sclerosis since 2016,' Hannah says. 'We were the first family they worked with. Because of our success, they are now helping other families of children with treatment-resistant epilepsy who need their very sick children to have access to medical cannabis in the UK.'

End Our Pain put Hannah up as a spokesperson. The resulting campaign was high-profile and quickly gathered momentum, helped by the fact that Alfie's story was so emotive.

'After my first media interview, on *BBC Breakfast*, the Home Office got in contact and said they would help Alfie,'

Hannah says. 'Then they put out a statement, saying cannabis had no medicinal value.'

'I had set up a petition with Change.org for Alfie. And after the first lot of media it got a flurry of signatures, went up to 365,000. I went to Downing Street, to present it to Number 10. My family and I were invited in for tea with Nick Hurd, minister of firearms and policing, who was in charge of cannabis because it was a Schedule 1 drug. Alfie wasn't well, he was climbing all over the sofas, so it wasn't easy.

'The door opened, and in walked Theresa May. She told me she'd seen my interviews. And she said she'd allow Alfie's doctors to apply for a Schedule 1 licence to prescribe Bedrolite – Alfie's medicine from the Netherlands – on the NHS. That felt like a massive win for us. A Schedule 1 licence had never been granted for an individual before, only for pharmaceutical company research.

'So I agreed not to seek media attention, while the Home Office did the paperwork. But months went past, and it felt as if the Home Office were dragging it out.'

Billy Caldwell, then 12, was another child who'd seen his life transformed by taking cannabis oil, this time from Canada. So much so that, back home in Northern Ireland, his GP agreed to prescribe it. But local medical authorities had threatened the GP with a charge of gross medical misconduct if he continued to prescribe an 'illegal' drug.

With the hope of preventing Billy's condition deteriorating, Billy and his mum Charlotte Caldwell went to Canada to pick up his cannabis oil. Flying back into Heathrow, Charlotte declared she was carrying a six-month supply of Billy's medicine. It was seized by customs. 'The customs officers were very conflicted about removing the medication from me,' Charlotte said at the resulting press conference. 'One of them had tears in his eyes.'[12]

Billy's seizures returned within a few days. He became extremely ill, and had to go into intensive care. There was a public outcry. After speaking to the doctors in charge of Billy's case, the then home secretary Sajid Javid intervened to release his medicine from customs as a medical emergency.

'I was so pleased for Charlotte and Billy,' Hannah says. 'But my son was poorly too. I was working with the government to try to do the right thing. I spoke to Nick Hurd, who said he'd look at the licence application. But again, they didn't move quickly.'

'So three days later, I went on the *Today* programme on Radio 4, where I was interviewed by John Humphrys. I said I'd met the prime minister, and she'd met Alfie, and I'd appealed to her directly.

'On air, I said that she'd told me they would find a way in which our clinicians could be issued with a Schedule 1 licence to give my son the medicine that he had in Holland. And that I'd believed her. But that had been three months ago.

'I also said all that we'd been put through since then, was bureaucracy – hurdles after hurdles after hurdles.

'Later that day, I got a phone call to say Alfie's doctors would be issued the first Schedule 1 licence to prescribe cannabis on the NHS.'

Javid announced this in the House of Commons. And, at the same time, he asked Professor Dame Sally Davies, Chief Medical Officer for England and Chief Medical Adviser to the UK government, to review the therapeutic and medicinal benefits of cannabis.[13]

Citing data from the US and Australia, Davies said there was 'conclusive evidence of the therapeutic benefit of cannabis-based medicinal products for certain medical conditions and reasonable evidence of therapeutic benefit in several other medical conditions.'[14]

She recommended what subsequently became law in November 2018: cannabis moved from Schedule 1 – which covers drugs that cannot be prescribed – to Schedule 2, where they can be prescribed but still with severe restrictions on prescribing and importation.

Finally cannabis was back to being a medicine, for the first time in almost 50 years. The legislation allowed specialist doctors to prescribe cannabis-based medicines, in theory for any condition. But while at first glance it looked as if the UK now had the most liberal cannabis legislation in Europe, as the next chapter explains in detail, it's still extremely hard to get a prescription.

CAN I GET MEDICAL CANNABIS?

THE VAST MAJORITY of people in the UK who want to use medical cannabis won't get an NHS prescription. This leaves them with a difficult choice.

There are doctors – usually in specialist private clinics – who prescribe it. But this is only available to those who can pay. It can cost up to £2,000 a month for medical-grade cannabis imported from the Netherlands.

That's one reason why 1.4 million people in the UK buy their medical cannabis on the black market. But this carries the risk of a criminal record and up to two years in prison. Even worse, if a carer buys it for a patient, they can be prosecuted for dealing, which has a maximum sentence of 14 years[1] as well as seizure of assets under the Proceeds of Crime Act.

You can't be certain of the quality of black market cannabis or the consistency of the supply. You may not be able to buy cannabis containing enough CBD, as most on the market is high-THC skunk. And there's an ethical consideration too: street cannabis funds gangs.

That's why we founded the research and pressure group Project Twenty21: to help people access affordable prescriptions for quality medical cannabis, taken under medical supervision. Twenty21 is also creating a database of patients who take cannabis for different conditions, which will be a source of real-world evidence (RWE) for medical cannabis.

To join, you do need the agreement of your GP so we can get good data on your condition and the other medicines you have taken for it. The medical consultation fee is capped at £150, and the average fee for medication per month is £150. This is not cheap but it's similar to what people pay on the black market – and much cheaper than Dutch medical-grade cannabis.

We can only do this because we've been able to source medical-grade cannabis at a greatly reduced price. Parents of children with epilepsy, who take high doses, were paying on average £1,700 a month; they now pay around £600 a month. That can be life-changing. We know of at least one parent who sold her house to pay for her daughter's treatment from Holland.[2]

Twenty21's second purpose is to collect clinical and medical data. We have brought together academic experts in specific treatment areas, in order to create observational studies for each diagnosis, to build a database evidence of dosage, what works and the negatives of treatments too.

So far, we have signed up over 1,000 people over eight areas: pain, anxiety, PTSD, multiple sclerosis, Tourette's, ADHD, adult epilepsy and cannabis dependence. We are planning to expand to more conditions, including adult epilepsy. The first paper using Twenty21 data was published in May 2021.[3]

PURE EXTRACT CANNABIS MEDICINES VS WHOLE PLANT EXTRACTS

As I described earlier, there are currently two types of medical cannabis products, which makes the situation pretty confusing.

The first is the whole plant medicine, as taken by Alfie and Billy. There is a high-CBD version, which is what Alfie and other children with epilepsy usually take. And there are versions with various THC:CBD ratios. All will contain the minor cannabinoids as well.

The second type is licensed drugs made from cannabis. These are purified extracts of THC and CBD in exact proportions, with none of the other plant chemicals cannabis contains. The advantage of this approach is that it follows our current big pharma medicines development model; that is, it produces standardised compounds that can then be patented, trialled, licensed and marketed. Doctors are much more used to prescribing this kind of medicine; NICE (the body that decides for the NHS if drugs are cost-effective) has agreed to fund them for certain conditions, and so there have been more prescriptions for them.

The alcohol-based mouth spray Sativex can be prescribed by doctors for spasticity in multiple sclerosis; and Epidyolex, pure CBD extract, only for two rare forms of epilepsy. Both drugs are manufactured by a British company called GW Pharmaceuticals. Sativex is in Schedule 4 of the 1971 Act, a much less controlled schedule than whole plant cannabis medicines, and Epidyolex is in Schedule 5, with very minor controls.

So why don't doctors just prescribe these drugs to more people? At the moment, as we see, they are only approved for very specific conditions. Bringing a drug to market is expensive – to the tune of millions of pounds, largely due to the cost of randomised controlled trials, or RCTs. In order to allow doctors to prescribe these drugs for more conditions, there would need to be RCTs for each condition.

In an RCT, the subjects are divided into two groups. One group receives the treatment, the other doesn't, but gets a placebo instead. And neither the participants nor the people running the trial know which group they are in. The purpose of this method is to reduce bias and cut out the placebo effect.

It took 30 years for GW Pharma to get these two cannabis drugs to market, and they can still only be prescribed in such a limited way. Even though the Sativex trial data looked promising for both pain and spasticity in MS, NICE decided it only met their cost-benefit criteria for spasticity, and so it can't be prescribed for pain in MS.

The trial of Epidyolex added it to existing medication, a benzodiazepine called clobazam, in children with the two rare childhood epilepsies Lennox–Gastaut syndrome and Dravet syndrome. NICE initially turned it down, in their draft guidance, on the grounds of cost. Patient groups have managed to get some anti-cancer and neurological drugs prescribed at costs above NICE's threshold, but only with some serious lobbying.[4]

Then, in November 2019, following the previous year's change of the legal status of cannabis and after some price negotiation, NICE revised its guidance and allowed both drugs to be prescribed on the NHS in England; although as we have seen, Epidyolex can only be prescribed for these two specific and rare childhood epilepsies and Sativex only for spasticity in MS.[5]

What the GW Pharma experience shows is that the traditional big pharma, drug development route to market for cannabis is both very slow and very expensive.

Also, patients often prefer whole plant extracts. For example, Sativex comes as a fixed THC:CBD dose and some MS patients have reported finding its THC component too strong.

Many patients have reported better symptom control for pain and inflammatory bowel disease from taking whole plant extracts, too. My group at Drug Science has recently shown that some children with treatment-resistant epilepsy report doing much better on the whole plant extract than they do on pure CBD Epidyolex.[6]

Cannabis is complicated, containing more than 100 molecules that work together in myriad ways. We don't yet know exactly how it happens, but we think they work in synergy, the entourage effect that we have looked at. Better symptom control could be due to this, or it could be that there is another component of the plant, aside from THC and CBD, that's helping.

Doctors who prescribe cannabis usually prescribe on a start low, go slow basis, beginning in most conditions with a high-CBD version. That allows them to find the optimal combination of CBD, THC and the other active chemicals of the cannabis plant with minimal risk of THC intoxication. This will often differ for different disorders, and even between patients and over time.

Whole plant extract allows people this flexibility. Alfie Dingley, for example, recently had to switch to a higher-THC formulation in order to control his seizures.[7]

SO WHY CAN'T DOCTORS PRESCRIBE WHOLE PLANT EXTRACT?

In 2019, Professor Chris Whitty, chief scientific advisor at the Department of Health, told the Health and Social Care Committee in the House of Commons that expanding patients' access to cannabis as a medicine risked becoming another thalidomide scandal.

'It's very dangerous to have a cannabis exceptionalism here. These are drugs, they have side effects, they have positive effects – that is clear.

'What we have to do is balance those two, but they are no different to any other drug in that sense.

'And if you look at the history of medical development, history is littered with people rushing things through and ending up regretting it or, in a few cases – thalidomide probably the most well-known – having an absolute disaster on their hands.'

In theory, this sounds sensible. NICE and Whitty are basing their decision on the principle of the old Hippocratic teaching to doctors: 'First Do No Harm'.

But surely Whitty knows that cannabis is not like other drugs. It's a plant chemical that has been used by human beings for thousands of years, not a new drug. It's already used as a medicine in 30 countries. Both THC and CBD have been through animal tests showing they do not cause foetal abnormalities – why should he want more animals put through this process? There must have been millions of babies born to mothers who've taken cannabis.

We do need cannabis exceptionalism.

Cannabis can't currently be classified as a medicine in the UK. As a plant extract, it can't be standardised to the level that a medicine requires.

But we can get over this hurdle, and manufacture it as a safe and consistent plant medicine. In Germany, for example, licensed herbs are classed as medicines and paid for by health insurers. And obviously there already exist standardised medical cannabis products that have reached a quality control standard, such as those taken by Alfie Dingley, imported from the Netherlands, and Billy Caldwell from Canada. In the Netherlands, there's an Office for Medical Cannabis, a government body responsible for regulating and supplying medical cannabis. It recognises that cannabis is different from other medicines and needs to be treated as such.

The point is, medical cannabis is extraordinarily safe, especially compared to a lot of other medications. We know CBD protects against some of the less desirable effects of THC, such as dependence, so it has safety built in. And we know the benefits for some patients are very high.

Modern medicine is about net benefit–risk, not just harm avoidance. Otherwise, since any treatment can have risks, we'd have to eliminate almost all medicines and surgery.

Red tape

In 2018, as we have seen, cannabis medicine was moved from Schedule 1 to Schedule 2 of the 1971 Act, the same as heroin and methadone. The result is, each patient needs their own separate import licence. This red tape adds time and expense and more potential for hold-ups. And, as some of the families of children with epilepsy have discovered, the situation is even more precarious post-Brexit.

There's red tape for doctors, too. Because cannabis is Schedule 2, doctors have to register as a prescriber plus apply for a special pink prescription pad just for cannabis.

With the exception of Sativex and Epidyolex for very limited conditions (plus an older medicine called nabilone for nausea during chemotherapy), the rules say cannabis medicines need to be prescribed as a 'special', also called off licence.

The truth is doctors prescribe off licence all the time. But many are reluctant to prescribe cannabis, for fear they won't be covered by their insurance.

They should be confident they are covered, in particular for the many indications where cannabis does have good evidence (see Chapter Ten), and by the fact that it has such a good safety profile.

There's a radical solution that could remove most of the red tape and help doctors to know that cannabis is safe. And that is for the government to move all whole plant medical cannabis products to Schedule 4 of the 1971 Act. CBD products – because they are non-psychoactive – should be removed from the Act and Schedules altogether. Since 2020 Drug Science has asked the ACMD to consider this several times but at the time of writing they have not replied.

Some doctors don't agree with cannabis

A lot of doctors are still uncertain about – or even against – prescribing cannabis. Doctors aren't taught about cannabis in medical school and, not surprisingly, they're reluctant to prescribe what they don't know and haven't experienced. Especially when you put that in the context of years of propaganda casting cannabis as dangerous with no medical value. If doctors have learned about cannabis at all, it will have likely been only in the context of it being illegal and the harms, such as dependence.

But we know that people asking for prescriptions are doing so because they want to feel better, not because they want to get high. In truth, there's little crossover between recreational and medical users.[8]

Currently, only specialist doctors are allowed to prescribe cannabis medicines; but getting to see one of these specialists can take months, if not years. GPs can only take over prescribing once a specialist has given the first prescription. But cannabis has a lot of potential value for conditions commonly seen and treated by GPs, such as anxiety, pain and sleep issues.

Another issue is that specialists are often invested in their own research interests or trials, rather than in medical cannabis. It's the patients who are bringing this treatment to the doctors, and unfortunately some doctors – particularly specialists – can be resistant to the idea that patients know best what works for them.

The British Paediatric Neurology Association have said that, for children with treatment-resistant epilepsy, brain surgery should be tried before cannabis. Understandably though, parents would prefer their children to try cannabis before undergoing brain surgery. The doctors' position here reflects an almost religious hostility to cannabis.

Who pays for it?

Even if a doctor wants to write a prescription for a patient, it's not clear where the money might come from.

One reason is the lack of NICE guidance on whole plant medicines.

When NICE wrote their guidelines, they had to be seen to be giving some concessions both to the parents of the children with epilepsy and what the government – the Chief Medical Officer – had said. So they did not make a recommendation against the use of whole plant medicine but they did say there wasn't enough evidence to make a recommendation in favour of them.[9]

But by doing this, NICE effectively denied them to patients, because doctors usually rely on NICE to guide what they prescribe – even though NICE is really about money.

Recently, NICE tried to fix this with a position statement on treatment-resistant epilepsy, empowering doctors to make the clinical decisions. But this didn't go far enough.

Even if an NHS doctor did want to prescribe whole plant medicine, they'd have to make a special case to their local Clinical Commissioning Group (CCG) to get the funding. And so far CCGs have almost always said no because NICE hasn't said yes. Parents of children with epilepsy are caught in an awful double-bind and their children are suffering terribly.

The wrong kind of evidence

Everyone – Whitty, most doctors, NICE – is asking for time to gather more evidence on cannabis.

What they mean by evidence is Randomised Controlled Trials (RCTs). Over the past 50 years, this has become the main way that drugs are tested. But the lack of RCTs doesn't mean medical cannabis doesn't work. It just means companies haven't conducted and may never conduct RCTs to prove it does work. These are some reasons why:

* **The trials would be too expensive**
  As I explained earlier, RCTs are prohibitively expensive. Running RCTs for all the different conditions for which people are taking cannabis and all the different dosages and combinations would take decades and cost hundreds of millions of pounds.
* **Cannabis is too complicated**
  RCTs really only work for simple, single-molecule pharmaceutical drugs, not complicated plants. If you included only high and low strengths of THC and CBD, that would be four three-arm studies (including placebo), which would require 12 groups of patients. And that's without including all the possible strengths and all the other active chemicals in cannabis and the fact that often people will need to adjust their dose and type of cannabis to optimise their treatment.
* **Companies can't patent a plant extract**
  Drugs companies won't do RCTs if they can't make their money back. It's not clear that a drugs company

could patent a plant medicine in a way that would protect their investment, or that NICE are willing to pay the price it will cost to bring one to market. And given that an alternative is already available on the black market, the chances of them recouping their investment look very small. Or a company could spend millions of pounds on trials, only for another company to come along with a similar and cheaper product.

* **We can't do RCTs for all patients**

So many patients have rare conditions that won't be covered by RCTs. How could you fund or do a trial on Alfie Dingley's condition, when he is one of only nine children in the world who has it? We cannot make patients wait for long and costly trials. Cannabis transformed the life of Lucy Stafford (see Chapter Eleven) and yet her condition is so complicated and rare, it's unlikely any company will ever fund a study. She could have continued going in and out of hospital, and likely died waiting for a study that was never going to happen.

The nature of evidence

There is a more fundamental point here too, around the nature of medical evidence. RCTs are not the only kind of evidence doctors and regulators should count.

We have extensive real-world evidence for medical cannabis, from databases of people who've taken medical cannabis in other countries, as well as one in the UK (see below).

In 2008, the then chair of NICE, Professor Sir Michael Rawlins, delivered a prestigious speech called the Harveian Oration, in which he pointed out that RCTs are too expensive and take too much time. He cited one drug trial that cost £95 million! But he also noted that RCTs are done on 'pure' patients, excluding those with other conditions, and most real patients have multiple conditions. We've found this with Twenty21 – most of our patients wouldn't be accepted on to a RCT for this reason.

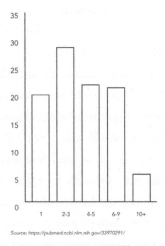

Source: https://pubmed.ncbi.nlm.nih.gov/33970291/

Fig. 7. Per cent of people in Twenty21 with different co-morbid conditions

Rawlins said that the kind of real-world evidence we do have for cannabis is better at helping reveal a drug's benefits and harms. And that if treatment produces an obvious benefit – as in the millions who've been treated with medical cannabis – RCTs become unnecessary.

In reality, doctors can't even predict if a drug that's undergone an RCT will work on a specific person. By the time patients come to cannabis, for example, they've usually tried and failed with the medications proven by RCTs. It may not have worked, or the side effects were too disabling or the risks were too high.

We already have a precedent for not relying on RCTs: there are 57 medicines currently available in Britain that are approved by NICE that have not undergone RCTs, most for cancer.[10]

In 2021, NICE put out a document saying that they would, in future, look at other sources of evidence, not just RCTs. Time will tell whether their actions match these words.

A new kind of medicine

So far, development of medical cannabis has been driven by patients. Doctors, science, industry and policy-makers are playing catch-up. We've never before had patients taking drug development into their own hands.

Medical cannabis is a game-changer, forcing us to completely rethink how we develop medicines in the UK. Doctors are now being asked to work together with patients to learn a whole new branch of pharmacology.

I think, and many colleagues agree, that this is a very rewarding thing to do. It puts the individual patient back at the centre of medical decision-making, rather than decisions on treatment being made by an algorithm of cost-benefit based on clinical trials involving other people.

We do need more research to find out about the safety, efficacy and side effects of cannabis. We do need to know more about short- and long-term health effects, both harms and benefits. But with careful monitoring, we can do this with real patients, in real time.

The government now needs to change the rules on cannabis medicines. Moving them to Schedule 4 and setting up a body to deal with cannabis medicine would be a good start.

For the government to do nothing now is unethical. It's driving people to the black market, or leaving them to suffer. This is exactly the opposite of what medicine is about. Medicine is about reducing suffering, not about proving a drug or intervention is better than placebo.

My sister, epilepsy and me

Chelsea Leyland is 33, and her sister Tamsin is 37. Both have epilepsy. Chelsea, a DJ and entrepreneur, has been using whole plant extract and CBD to control her seizures for five years. These are her words.

Tamsin has a much more severe form of epilepsy than me, known as refractory epilepsy. This is also called treatment-resistant epilepsy and it means no anti-convulsant has been able to keep her seizures completely under control.

She started having seizures as a baby, often every day. When her condition became too hard for my parents to manage, when she was seven and I was two, she had to move away from home. She now lives at the Epilepsy Society home in Chalfont St Peter.

Over the years, Tamsin has probably tried every anti-convulsant drug that exists. And she was one of the first patients in the UK to have a vagus nerve stimulator, a small electrical device, put into her chest, which has probably saved her life.

She still has many different types of seizures. The scariest kind is the drop attack, or atonic seizure, where she drops to the ground with no warning. Or she can be eating lunch and end up in it face first. Every time she falls, she cracks her head and has cut her head open numerous times.

CANNABIS

When Tamsin was little, doctors said she wouldn't live past the age of 18. So she's doing pretty well. I don't know how she does it, she takes such a beating on a daily basis. She often needs to sleep most of the day because of the impact of the seizures, and the anti-convulsants make her lethargic and sluggish. She can be so overly sedated that she'll dribble while she's eating.

Tamsin is such a sweet-natured, delightful, kind and funny person. I know she'd love to be living a different life. She sometimes says, 'I wish God hadn't given me epilepsy.'

She's able to come home for some weekends, for a week at the very most. She has to be watched all the time. On a good day we can sit in the kitchen, have a conversation. On a very good day, we can even go shopping. On a bad day, she sleeps and has seizures.

My epilepsy is milder. I was in my teens when I first noticed symptoms. In the car, when we drove through areas with dappled light from the trees, I'd get a funny sensation. Then I began to get what's called myoclonic jerks in the morning. Historically, this was called flying saucer epilepsy because whatever you're holding at the time – cup of tea, cereal bowl – your hands open and it falls out.

My diagnosis was a big shock for my parents. They'd already had such a struggle with Tamsin, and I had always been the cognitively healthy child.

At first, I didn't want to take any medication, I'd seen what it had done to Tamsin. But aged 15, I had my first tonic-clonic seizure, where you lose consciousness. It was incredibly frightening. After that, the sensible thing to do was to take medication, because of the damage seizures can cause in your brain.

The neurologists' focus is on finding medication that eliminates or at least reduces the seizures. Quality of life comes second to that.

I started with a milder anti-convulsant, Lamictal [lamotrigine]. It worked temporarily, although I still had seizures on this drug, so the neurologist then added in Keppra [levetiracetam], a more powerful medication.

Keppra made my seizures manageable and eventually I could drop the Lamictal. But I found the Keppra side effects terrible. I was riddled with anxiety, felt depressed and was hyperactive at night. It's a terrible irony: sleep deprivation is a trigger for epilepsy, but the meds keep you awake. I had a terrible temper, would fly into a rage. I've now started a women's epilepsy support group and I've heard stories from other women about their anger on this drug too. One woman stabbed her boyfriend with a fork. But at the time, I couldn't tell what was me and what was the medication. I had to take anti-anxiety medication and sleeping pills, just to function.

As well as epilepsy, I also had endometriosis and linked gut issues. Every month during my period, I'd need a lot of painkillers – mefenamic acid and codeine. The pain was sometimes so bad I'd pass out, vomit and have to go to A&E. I had laparoscopic surgery at 26 – but I was still regularly in pain.

Looking back, it's surprising that I managed to hold down my job as a DJ. For ten years, I spent a lot of my life on a plane, flying around the world. I did lose a few jobs due to having epilepsy. Sometimes it was down to not being able to perform around strobe lighting and some people don't want the risk of hiring a DJ with epilepsy.

Then, five years ago, I was living in the US, and I was introduced to CBD. Even the first time I sprayed the CBD tincture under my tongue, I noticed something quite profound was happening. I remember driving home that night, at 2 a.m. I wasn't high but I had a feeling of ground-edness, calm and ease, the opposite to my epilepsy, which feels wired.

That night, I forgot to take my Keppra, which had never happened before. This might not sound like a big deal but I was so petrified of having a seizure that I was more likely to take it twice than not at all.

The next morning, I realised why I'd forgotten. Usually, my brain would start to feel strange in the evenings, and that would remind me to take it. But the night before, the CBD had made me feel as if I'd taken it.

A few years before, I had watched the Charlotte's Web documentary on CBD and thought, this cannot be real. Growing up with Tamsin, I couldn't imagine that a plant oil could be powerful enough to stop seizures.

But taking CBD, I could feel the difference. I started taking it every day. I felt so good that, over four months, I began to cut down on Keppra. (I wouldn't advise anyone to do this without the help of their doctor).

After I finally stopped taking Keppra, I felt as if I was getting to know the real me again. I could think and breathe, I could retain information, my emotions were calmer.

By this time, I was living in the US. Under medical advice, I began taking a 1:1 extract of THC:CBD as well as a full-spectrum CBD product.

When I first took THC, I could feel the psychoactive properties. But that high feeling went away after few months, so it left me chilled rather than stoned. I don't take THC every day. It's not great for my memory, and sometimes I don't want to take it or I can't get it, such as when I'm in the UK, and then I take CBD alone.

CBD makes me tired – it doesn't do that to everyone. I usually vape in bed, take two to five drags, yawn immediately then I'm asleep in 15 minutes. I went from hardly sleeping, to sleeping 12 hours a night! I was so grateful, it felt like a gift.

I haven't had a big seizure since I started taking cannabinoids. I used to get a cluster of myoclonic jerks every morning, often followed by a seizure. But now I have perhaps a single one every four to five months. I realised how well it was working for me, when I was offered a DJ job in Bali, but decided to turn it down. CBD is illegal there and I didn't want to stop taking it, even for such an amazing work trip.

My endometriosis pain has also improved dramatically and particularly since I started using cannabinoids intra-vaginally. This was the catalyst for the women's health business I co-founded. Our hero product is Looni's intra-vaginal suppository using a specific combination of cannabinoid compounds and terpenes to help with acute menstrual pain, which is coming to market early next year. I start using them two to three days before my period is due. A tiny bit of THC, vaped, is really helpful when the pain is bad, too.

For eight years, my family and I have been asking Tamsin's doctors if she can try medical cannabis.

They have told me: Tamsin isn't responsive to anti-convulsants so why would she be responsive to this? That's despite cannabis working in a completely different way to anti-convulsants.

We've been told there isn't enough medical research that proves cannabis works. It's true: but there is an abundance of anecdotal evidence and the science hasn't caught up.

We took Tamsin to see a specialist in cannabis medicine and epilepsy for an opinion, who thought it was worth trying. And a neurologist I saw in the US said the fact Tamsin and I have the same genetic mutation suggests it may be a good treatment. But her neurologist will not agree.

All we are asking for is a trial. We are not expecting a miracle. We know the degree of Tamsin's condition and of her brain damage, too. But if it could improve Tamsin's sleep, or just mean she could cut the dosage of her anti-convulsants, improving her quality of life, that would be amazing.

I have had the most incredible life transformation due to cannabis. It seems so unfair that if Tamsin was as well as I am, she could make her own decision about CBD and cannabis. But she is vulnerable, she can't fight for herself, and so she can't have access.

Rather than focus on what I can't do for my sister, I focus on what I can do. I am running a support group for female epilepsy patients. And I've made a documentary about

Tamsin and my experience, called *Sisters Interrupted*, which is currently in post-production.

There's such a big change around cannabis medicine going on in the US, I have to believe there will be a change in the UK in Tamsin's lifetime too.

10

THE PROS AND CONS OF CANNABIS MEDICINES

CANNABIS ISN'T REALLY one medicine but a whole family of medicines.[1]

It works on the endocannabinoid system, which balances all the body's other systems, so it has wide-ranging effects. It's anti-spastic, analgesic, anti-emetic, neuroprotective and anti-inflammatory.

It's not a panacea or a miracle treatment but it's certainly a treatment worth considering for people with a whole range of conditions. That said, there is a lot more evidence for some than for others.

Research so far shows that getting the proportions of THC and CBD right seems to be key for specific conditions.

It seems that some of the other cannabinoids will turn out to be important, although here research so far has been extremely limited. And it might be that it's the entourage effect of the whole plant that turns out to be the most powerful of all.

This is an exciting area of research, one that's likely to change dramatically in the next decade. If you are thinking of taking medical cannabis, I would always advise you see a doctor with some expertise in the field (find one via drugscience.org.uk). Taking street cannabis could be risky and is much less likely to give you the results you want. This chapter will explain the evidence so far, and hopefully guide you to knowing if cannabis is for you.

COMMON QUESTIONS I'M ASKED ABOUT MEDICAL CANNABIS

IS IT POSSIBLE TO FATALLY OVERDOSE?

No. Cannabis is extremely safe. The amount of cannabis it takes to kill you is over 1,000 times an average dose. As a comparison, the amount of opiates or alcohol that it takes to kill you is just two to four times an average dose.[2] There are no proven instances of someone dying from a cannabis overdose – although it's possible someone with bad heart disease might have a heart attack due to the increased blood pressure. You could say that, for most people, the only way you can die from cannabis is to be crushed by a bale of it![3]

It's likely that skunk – the strong cannabis that makes up most of what's sold on the illicit market now – is less safe as it doesn't have the protective effect of CBD.

One warning: never use any synthetic cannabinoids, also known as spice and Black Mamba (see Chapter Thirteen for more). These are NOT the same as plant cannabis. They are much more powerful, have unpredictable effects and can cause severe mental disruption, seizures and heart attacks.

CAN YOU GET ADDICTED TO MEDICAL CANNABIS?

It's unlikely. The usual statistic is that 8 to 10 per cent of people who try cannabis become dependent. However, using the terms addicted or dependent about people with medical needs is misleading and even stigmatising. People who need insulin for diabetes are not addicts even though they are dependent on the medicine to stay well. The same is true of people who are prescribed opiate painkillers for pain.

Secondly, if we were to measure dependence in medical users in the same way as non-medical users, it's likely the percentage would be much lower. Most recreational users use high-strength cannabis or skunk, which we know is more addictive. Medical users use high CBD or balanced CBD:THC mixtures, and under medical supervision. And the preferred way to take medical cannabis is as an oil, which gives a delayed and lower 'high' compared to smoking. This is a major factor in reducing the risk because a faster and higher high increases the risk of becoming dependent (see Chapter Twelve for more). Finally, medical users titrate their dose to deal with their symptoms and not in order to get high.

WHAT ARE THE DOWNSIDES OF MEDICAL CANNABIS?

It won't suit everyone. Like all medications, it has risks and side effects. You may find you don't like the feeling of being high that you get from THC (CBD doesn't make you high). For example, you may feel dizzy or lose your balance, hallucinate or get blurred vision or mood changes. But these are all short-term changes. That said, if you feel under the influence and you have taken THC it's advisable not to drive (see Chapter Fifteen for more detail).

You may have heard there are longer-term risks, such as mood disorders, but this is unproven and not clear-cut. Smoking cannabis may increase your risk of cancer. See Chapter Thirteen for more on the possible long-term harms.

If the advice from a patient charity that covers your particular condition is negative about medical cannabis, it's worth knowing that this kind of organisation tends to be cautious. They are often staffed by law-abiding citizens who are worried about encouraging the use of an illegal drug. The charity will want to protect itself against any claim of recklessness. And their view will often also reflect that of the medical establishment; which, as illustrated in the previous chapter, is still relatively hostile to medical cannabis.

You need advice based on current knowledge that's specific to your condition. The best way to get this is to book into a clinic that specialises in medical cannabis and discuss your condition with a doctor there. Find your nearest clinic under the Project Twenty21 tab at drugscience.org.uk.

ARE THERE ANY CONDITIONS THAT MEAN I SHOULDN'T TRY CANNABIS?

I know of no medical condition that is an absolute contra-indication for medical cannabis. There are relative contra-indications, i.e. conditions that might be aggravated particularly by high-THC products. These include cardiovascular disease, hypertension, bronchitis and asthma (particularly if smoked) and psychosis. For all of these, it's better to take medical cannabis as an oil rather than vaporising, where possible.

Current Food Standards Agency (FSA) guidance says pregnant and breastfeeding women should not take CBD and I would advise the same for cannabis. The FSA also states you shouldn't take CBD if you are taking other medications (see below for more on this).[4]

DOES CANNABIS INTERACT WITH OTHER MEDICATIONS?

Again, you'll need advice from a doctor. Cannabis can interact with other medications in two ways. Firstly, its effects can add to the effects of a medication you're taking. For example, if you are sleepy from taking sedatives it might make you more sleepy.

Secondly, it can alter the breakdown in your body of other medicines, speeding it up or slowing it down.

This is an area with a lot of theory but very little data. CBD is known to block or slow the metabolism of certain medicines that are broken down in the liver.

If you are taking the following drugs, talk to your doctor: warfarin, amiodarone (a heart rhythm medication), levothyroxine (a thyroid medication), valproate (used for epilepsy, bipolar disorder, migraine). In general, the risk is greater with CBD because people, especially those with epilepsy, take much higher doses of this than the CBD/THC mixtures.

A Harvard review states the following medications may have interactions with CBD: opioids, benzodiazepines, antipsychotics, antihistamines, sedating antidepressants such as amitriptyline.[5]

CBD has been shown to interact with other drugs, including clobazam. That said, in an observational trial of people with epilepsy who were taking CBD with a whole range of other anti-epilepsy drugs, no obvious problems emerged.[6]

High doses of CBD can lead to changes in liver function, as you make more enzymes to process the CBD. Usually this is transient. And it is usually only an issue in epilepsy patients, who can require up to 20 times the average dose, and because these patients are often on other medications that can also irritate the liver.

DO I NEED CBD OR THC OR BOTH?

It depends. The advice when taking medical cannabis is to start low, go slow. A doctor will almost always start you on CBD oil, as it has fewer side effects. For well-being, a suggested starting dose of CBD is 10mg to 30mg a day. For health conditions, it's 50mg. However, this will vary.

You may have read warnings that CBD oil, usually extracted from the hemp plant, contains THC. Even the 'pure' CBD drug Epidyolex contains 0.3 per cent THC, which is 0.3mg per 100mg of CBD. However, this amount of THC, like the small amount in CBD oil, is so small that it's very unlikely to have any noticeable effect.

If CBD alone doesn't help, or doesn't help enough, the next step may be for your doctor to prescribe a medication that does contain an active amount of THC, usually starting with a CBD-dominant strain of cannabis. Some conditions – for example severe pain – may respond better to a product that contains mostly THC, though again some CBD might help balance out any negative effects.

NB: To maximise the effects, take your CBD or medical cannabis with a fatty meal, as that helps with absorption.

IS IT EVER SAFE TO BUY IT ON THE ILLICIT MARKET?

The short answer is no. If you can't get an NHS prescription, you could try one of the registered private clinics that specialise in cannabis prescribing. Or a subsidised prescription from Twenty21, which is run by my charity Drug Science. You do need to have one of the conditions we are currently researching: anxiety disorder, chronic pain, multiple sclerosis, PTSD, substance use disorder, Tourette's syndrome, adult epilepsy. See drugscience.org.uk for more information.

One big issue with street cannabis is that you likely won't know what's in it. It varies in strength and make-up, and is usually much higher in THC than medical cannabis. It can also be full of impurities.

The other risk is of being arrested. One important consideration here is the attitude of your local chief constable and crime commissioner. Some are relaxed, some are not. Chief Constable Mike Barton, when head of Durham Constabulary, wrote that police in his area would arrest someone who's smoking cannabis on the street, but not who's smoking at home or growing a single plant for their own use. 'We're not going soft on drugs, we're going sensible,' he wrote. Other forces take a very different view.

The situation is more problematic if someone else is sourcing and purchasing cannabis for you. Under the law, this is dealing and it can incur a 14-year prison sentence. But worse – if the police choose they can prosecute people for dealing under the Proceeds of Crime Act, freeze their assets and cause them an enormous amount of grief. This law can also be used to prosecute people who are growing to supply others for their health needs.

If you are not able to source medical cannabis legally, you can apply for a Cancard (at cancard.co.uk). It acts as proof that you have one of the conditions for which medical cannabis is a treatment. The website says: 'Police forces in the UK have been briefed on Cancard by the National Police Chiefs counsel that police officers should be confident in using their discretion when someone has small quantities of cannabis for medicinal purposes.' This scheme has influential supporters, including police and MPs.[7]

HOW MUCH WILL MEDICAL CANNABIS COST?

This really depends on what and how much you're taking. In particular, high doses of CBD can be extremely expensive, thousands of pounds a month. At Twenty21, we have capped the prescription price of a standard product at £150 a month. (There are concessions if you'd struggle to pay this.)

HOW DO I KNOW THE CBD I'M
BUYING IS GOOD QUALITY?

CBD is the one cannabis product that the UK has dealt with in a rational way. It wasn't included in the Misuse of Drugs Act 1971 because our scientists told the government it was not psychoactive. Most of the rest of the world assumed it was like THC and banned it.

However, there was very little interest in CBD until recently. One reason was that the source, i.e. the hemp plant, was illegal to grow unless you had a special licence. Now it's possible to make pure synthetic CBD, which avoids this problem.

However, almost all the CBD sold in the UK today is extracted from plants and will contain a tiny amount of THC, typically less than 1 per cent. That means if you are taking 10mg of CBD, you are also taking 0.1mg of THC. You are unlikely to notice any effects from this amount.

Legally, CBD is classed as a novel food, so it's covered by Food Standards Agency regulation. The marketing of CBD has recently come under scrutiny. Since 1 April 2021 all products sold in England, Wales and Northern Ireland have had to comply with FSA regulations and be approved. In Scotland, CBD comes under Food Standards Scotland. To be approved, companies need to show toxicological and safety checks, so it introduces a safety standard.

Tests in 2019 by the CMC (Centre for Medicinal Cannabis) showed that only 38 per cent of the products were within 10 per cent of the advertised CBD content, and the same number actually had less than half of the advertised CBD content. One product even contained no CBD at all. Do your research and go for the best quality you can find.[8]

DO I NEED TO TELL MY GP I'M
TAKING CBD OR CANNABIS?

Yes, it would be sensible to tell your GP. If you go to Twenty21 or any other medical cannabis prescribing group, they will require a note from your GP that you do suffer from the relevant indications and therefore are eligible for the treatment.

As part of the application for a Cancard, you have to let your GP know you're taking cannabis. But you might want to feed back the results to them too: this has the great advantage of educating them about the benefits.

WHERE CAN I FIND OTHER PEOPLE WHO
ARE TAKING MEDICAL CANNABIS?

PLEA or Patient-Led Engagement for Access is a non-profit campaigning for access to medical cannabis. End Our Pain is a campaign supporting people who are being denied NHS prescriptions. United Patients Alliance is another patient advocacy group (healtheuropa.eu) that's partnering with Project Twenty21. PlantEd Collective is an educational organisation aimed at women.

WHO IS USING CANNABIS AND WHY?

I have mentioned the survey by Dr Daniel Couch at the CMC that revealed over 1.4 million people in Britain are using cannabis to relieve symptoms of diagnosed medical conditions. This is more than 2 per cent of the population. The most frequent conditions cited were depression, anxiety, chronic pain, arthritis, PTSD, autism spectrum disorder (ASD) and inflammatory bowel disease (IBD).

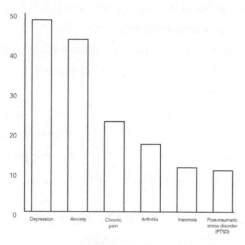

Source: https://thecmcuk.org/news/million-uk-adults-self-medicating-with-illicit-cannabis

Fig. 8 Top 6 reasons people use medical cannabis

Further evidence comes from databases created in other countries and territories where medical cannabis is legal. In Germany, 71 per cent of prescriptions for cannabis are for chronic pain. In the US state of Minnesota's medical cannabis patient registry database, 62 per cent are using it for pain. In the Canadian database, 30 per cent are using it for anxiety.

Our first evidence from Twenty21 is showing interesting results on quality of life. We have discovered that most of the people getting treated have pain or anxiety syndromes as well as one or more other medical conditions, so are quite disabled. Their quality of life comes out at around half that of the general population. However, the first outcome data shows that after just three months of treatment, most patients reported their quality of life increased significantly.[9]

In the next chapter, we'll look in detail at specific conditions and the evidence so far. While medical cannabis isn't a perfect treatment, real-world evidence suggests it may be improving quality of life for many millions of people.

11

WILL CANNABIS WORK FOR ME?

MILLIONS OF PEOPLE are using cannabis for multiple conditions, but there's a glaring lack of research. For some conditions, cannabis is a well-established treatment. For others, it's still controversial.

Often, it's conditions with small patient numbers that have the strongest empirical data. And the conditions that affect the most people often have debatable clinical evidence but good RWE (Real World Evidence).[1]

In order to help you make a decision as to whether cannabis is right for you, this is an assessment of the evidence that exists at the moment, although this is an area where change is happening fast.

The conditions are divided into five groups, reflecting the

amount, quality and reliability of data. Group 1 has the most evidence, and Group 5 the least.

HOW GOOD IS THE EVIDENCE?

1 High strength of evidence, high number of patients
   - Neuropathic pain
   - Cancer-related pain
   - Chronic pain

2 High strength of evidence, low number of patients
   - Intractable epilepsy
   - Multiple sclerosis
   - Cancer-related nausea
   - Appetite stimulation in wasting disorders
   - Inflammatory bowel disorders
   - Ehlers–Danlos syndrome (EDS)

3 Low strength of evidence, high number of patients
   - Arthritis
   - Sleep disturbances
   - Anxiety
   - Depression

4 Low strength of evidence, low number of patients
   - Parkinson's disease
   - Tourette's syndrome
   - Skin conditions
   - Fibromyalgia

- Glaucoma
- Dementia
- Autism spectrum disorder (ASD)
- ADHD
- Substance use disorder (SUD)
- PTSD
- Migraine
- Schizophrenia

5 Other conditions – for which people are using cannabis, but which have no or very limited evidence
- High blood pressure
- Cancer
- Immunosuppression and/or HIV/AIDS
- Huntington's disease

PAIN

Pain is the most common reason people use cannabis, some accounting for up to 90 per cent of users, according to state-level medical cannabis registries.[2]

This isn't surprising: an estimated one in five adults has chronic pain.

Pain is also the condition with the most evidence in terms of trials. There is good evidence for neuropathic pain (chronic pain from damaged tissue), and growing evidence for other forms of chronic pain, for example arthritis and cancer-related pain. And, as you'll see below, people use cannabis to treat pain in many conditions, including arthritis but also multiple sclerosis. Its benefits may come from actions apart from pain relief, such as being anti-inflammatory.

However, at present, the only guidance from NICE to doctors on prescribing cannabis for pain is that they can continue if the patient already has a prescription. This is extraordinarily conservative, given the vast amount of real-world data from patients that shows it works. And it ignores the fact that people in pain are, effectively, voting with their feet by using illegal cannabis.

In 2021, Drug Science rated the treatments for chronic neuropathic pain in a systematic way, using multi-criteria decision analysis. For the study, we gathered an expert group of pain specialists, psychiatrists and patient experts to assess 12 widely used pain treatments on 17 separate safety and benefit criteria.

We included three kinds of medical cannabis: THC/CBD at a ratio of 1:1, CBD-dominant and THC-dominant.

The other treatments included duloxetine, the gabapentinoids, amitriptyline, opioids such as tramadol, fentanyl, morphine and oxycodone, and ibuprofen.

The purpose was to rate products on their efficacy in pain control and the quality of life they give people. We also assessed their unwanted and adverse effects, such as cognitive impairment, constipation, dizziness, drowsiness, withdrawal and mood impairment, risk of dependence and risk of death.

Cannabis was not the best pain reliever per se, in that it didn't cut pain as much as, for example, the gabapentinoids. However, the group gave more weight to quality of life than

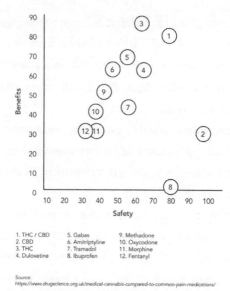

Source:
https://www.drugscience.org.uk/medical-cannabis-compared-to-common-pain-medications/

**Fig. 9 Safety vs benefits comparison for medicines used
in people with chronic neuropathic pain[3]**

pain relief alone. After all, that is the whole purpose of medicine. And that made cannabis a significantly better and preferred option than other pain drugs. Opiates, by comparison, rated very badly on mental effects, safety and risk of dependence.

Patients preferred cannabis as it had less in the way of unwanted side effects. THC alone and THC:CBD-balanced mixtures performed better than CBD alone. But CBD came out as the safest drug.

This may explain why cannabis is so widely used for pain relief: it works but it also allows people to live.

This is borne out by surveys of people who use cannabis for other pain conditions, for example, MS and arthritis (see below). And in an Australian survey of women with endometriosis, a difficult pain to manage, one in ten said they used cannabis successfully.[4]

EPILEPSY

Irish physician William Brooke O'Shaughnessy was a Victorian pioneer of medical cannabis, who published research of its effects on animals. But it was its effect on a 40-day-old baby girl with recurrent seizures that most impressed its usefulness on him, causing him to call it 'an anti-convulsive remedy of the greatest value'.

Cut to the modern day and, as explained in the previous chapter, both pure CBD and high-CBD–low-THC whole plant extracts are being used in the treatment of epilepsy.

In the UK, as we've seen, Epidyolex (pure CBD extract) is licensed for two rare kinds of childhood epilepsy, Dravet syndrome and Lennox–Gastaut syndrome, as an add-on to the benzodiazepine clobazam. However, there is growing evidence that some children may do better on whole plant extract CBD or CBD plus THC.[5]

There is some evidence in other rare forms of childhood epilepsy and in adults too.[6]

Adult epilepsy is a really high-priority area for research (see Chelsea's story, Chapter Nine).

In February 2021 the European Medicines Agency gave a positive approval of the use of Epidyolex for seizures in children with another rare condition: tuberous sclerosis complex. This is a disorder that presents in childhood, of benign tumours growing in many organs. When they grow in the brain, epilepsy is very common (in more than 90 per cent of cases). The condition is difficult to treat, with over 60 per cent failing to respond to standard treatments.[7]

My own view is that anyone whose epilepsy is not fully controlled should be given a trial of medical cannabis. There are over 300,000 people in Britain with chronic intractable epilepsy. And 600 people die of epilepsy every year.[8]

If cannabis helps just 10 per cent of people, this would be of massive benefit – to them, to their carers, to the NHS and to society.

Why does cannabis work? We don't yet know. It may be that people with epilepsy have a shortage of endogenous cannabinoids and CBD and/or THC can rectify this. One theory is that it increases levels of GABA, the main inhibitory brain chemical, normalising abnormal neuronal activity. Another is that CBD may help the brain neurons form new connections.

One of the minor cannabinoids – CBDV or cannabidivarin – has been trialled in adult epilepsy but didn't show great benefits. However, it may still be important as part of the entourage effect.[9]

MULTIPLE SCLEROSIS AND OTHER CONDITIONS THAT INVOLVE SPASTICITY

Both THC and a THC:CBD blend have been shown to be helpful, particularly for spasticity. After people with MS reported the benefits they experienced from smoking cannabis, GW Pharma developed the product Sativex, the mixture of purified THC and CBD that we have looked at. It is a solution of these two cannabinoids in alcohol (somewhat like Victorian cannabis tinctures). It comes as a mouth spray, so the cannabis is absorbed through the skin of the mouth and tongue.

Sativex also showed some effects on pain, but this did not reach NICE's cost–benefit threshold so is not available on the NHS for people with MS.[10]

There's some evidence that cannabis can help with spasticity in other conditions, such as spinal cord injury, too (see below).

CANCER-RELATED NAUSEA AND VOMITING

If you are feeling sick due to chemotherapy, you may be offered nabilone, a licensed THC medicine, as a treatment. THC has been used for this for over 30 years.

Some patients report that taking whole plant extract gives them significant benefits too, although there haven't been systematic studies of this, probably because of its previous illegal status and the fact that there are other, newer anti-emetic agents.

APPETITE STIMULATION IN WASTING DISORDERS

There's some evidence that cannabis and cannabinoids are useful for people with HIV wasting syndrome and chronic illnesses. But there is little evidence that cannabis can help with weight loss associated with cancer or anorexia nervosa.

There are some studies indicating that cannabis increases calorific intake. It may help you put on weight if you are underweight but also help you lose weight if you are overweight (see below). This further supports the idea that it can work as an adaptogen, balancing the body.

INFLAMMATORY BOWEL DISORDERS (IBD)

It's known that the endocannabinoid system is compromised in IBD. A lot of people use cannabis for both ulcerative colitis (UC) and Crohn's disease to reduce inflammation, pain and urgency. It's not available as an NHS treatment but it could help a lot of people; there are 300,000 people in the UK with these conditions.

In one survey, 25 per cent of respondents used cannabis for their IBD symptoms, including abdominal pain, stress, sleep problems, cramping and anxiety. And 93 per cent said it was effective.[11]

In two reviews of Crohn's and UC, taking cannabis didn't produce remission or a decrease in local inflammation. The smoked cannabis did reduce disease activity and both smoked and cannabis oil improved quality of life.[12]

PARKINSON'S DISEASE

Many patients say smoking cannabis helps with muscle spasms, pain and sleep issues, symptoms common in this condition. A significant number of Parkinson's patients self-medicate. There have been some small research studies; for instance a lab study showed that cannabis does improve tremor rigidity and movement. Interestingly, the study also reported that a large majority of doctors are comfortable with Parkinson's patients using cannabis, even though they say they couldn't recommend it.[13]

A survey showed that around half of patients who'd used cannabis said it helped their stiffness and shakiness, and that the effects happened over two months.[14]

SKIN CONDITIONS

Cannabis is a topical pain reliever as well as being anti-itch, antimicrobial and anti-inflammatory. Dr Henry Granger Piffard, one of the founders of American dermatology, wrote in the nineteenth century, 'a pill of cannabis indica at bedtime has at my hands sometimes afforded relief to the intolerable itching of eczema.'

There are endocannabinoid receptors in the skin, and these are involved in barrier regeneration.[15]

The dysregulation of the barrier is important in a lot of inflammatory skin conditions, including atopic dermatitis, psoriasis and acne.

More research is needed, but some clinics are already prescribing cannabis both internally and topically for conditions where existing therapies have failed, including acne. In a study on atopic dermatitis, a cannabis cream reduced itch and improved sleep by 60 per cent and a third of the patients were able to stop using steroid creams.[16]

And in a small observational study on both psoriasis and atopic dermatitis, a CBD cream improved damaged areas of skin. If cannabis can move people away from more harmful medications such as steroids, this could turn out to be a great advantage.[17]

FIBROMYALGIA

In this condition, people have chronic pain all over their body. They may have limited movement, disturbed sleep and often depression too. They can end up taking both painkillers and antidepressant medication. In an observational study at an Israeli pain clinic, after taking cannabis for six months, 80 per cent of people reported their pain had reduced, with ratings of pain going from nine out of ten to five out of ten.[18]

TOURETTE'S SYNDROME

The stereotype of someone with Tourette's syndrome is shouting swear words, but this neurological disorder is more complex. Tics can be spoken, but more often involve repetitive movements. The result can be muscle imbalance and pain from trying to restrain tics, as well as an emotional impact. Current treatments for Tourette's are not that effective and often have unwanted side effects. This is why we are now including Tourette's patients in Drug Science's Twenty21 initiative.

A case report describes a 12-year-old child with severe insomnia and tics, who was treated by his doctor parents with relatively high doses of THC, both vaporised and oral. His symptoms resolved.[19]

And a case review from Canada involving 20 patients with moderate to severe symptoms showed an average of a 60 per cent reduction in symptoms. Eighteen out of the nineteen were 'very much improved' or 'much improved'.[20]

DEMENTIA

There are no good systematic studies for dementia. But there is a lot of anecdotal evidence that cannabis may help to improve sleep and regulate night-time behaviour by stabilising circadian rhythms. It's now commonly used at night in Israeli care homes, for example. Anecdotal evidence also suggests CBD may also help with other behavioural symptoms of dementia, such as anxiety and agitation.

AUTISM SPECTRUM DISORDER

A person with ASD will have differences in sociability, learning and attention, which can come along with anxiety, aggression, panic, tantrums and self-harm.

There's often an increased prevalence of epilepsy in people with a diagnosis of autism; one theory is that the underlying aetiology of the two conditions may be similar.

There have been positive reports of treating autism symptoms with high CBD/low to moderate THC. In an Israeli case review, after six months of treatment, 84 per cent of children treated saw either significant or moderate improvement in symptoms. Most were taking cannabis oil containing 30 per cent CBD and 1.5 per cent THC.[21]

In an observational study of 18 children in Brazil, after six months, three had discontinued treatment due to adverse symptoms, but 14 saw an improvement in symptoms. The biggest improvements were seen in seizures, attention, sleep and communication.[22]

There aren't any medications licensed in the UK for autism symptoms, other than for aggression. So most medications given to people with autism will be unlicensed. Seen from that perspective, it doesn't seem so different to try other unlicensed medicines, namely CBD or CBD with a low amount of THC.

GLAUCOMA

This condition, increased pressure in the eye, can damage the nerves and is a common cause of blindness. It's been known since a study in 1971 that taking cannabis reduces intra-ocular pressure, by increasing the production and the outflow of eye fluids. However, this exciting initial finding has not translated into a clinical therapy, perhaps because the effects are short-term. So far there's no good evidence of long-term benefits.

DEPENDENCE

Patient experience shows people who start using medical cannabis for pain can reduce their use of opioid painkillers such as morphine, so reducing the risk of dependence to these medicines.[23]

It's been shown that CBD is protective against the risk of dependence to THC, so CBD and high-CBD/low-THC cannabis are being used in cannabis use disorder.[24] CBD may turn out to be useful to treat dependence to other substances too. For example, animal studies suggest CBD may reduce symptoms of alcohol withdrawal, as well as alcohol-seeking behaviour.[25]

And there are also preliminary studies showing it may be helpful in addiction to cocaine and methamphetamine too.[26]

Animal studies have shown that cannabis receptors link and feed back to moderate opiate and dopamine receptors, the targets of opiates and stimulants.

ARTHRITIS

Two of the most common forms of arthritis, rheumatoid arthritis and osteoarthritis, have major features that cannabis can help treat: chronic pain with its knock-on effect on sleep, and underlying inflammation. Animal studies suggest that CBD may help with both sets of symptoms, but studies on humans haven't yet backed this up.[27]

A 2006 GW Pharma study looked at the efficacy of Sativex on pain in rheumatoid arthritis; the effects weren't strong compared to other treatments.[28]

However, as shown in the pain section above, people are self-medicating using cannabis for arthritis pain because it gives them quality of life. As for other types of pain, best results may come from a mixture of THC and CBD. In arthritis, medical cannabis can be taken orally or as a cream or ointment rubbed into the affected joint to provide local relief without whole-body effects.

SLEEP DISTURBANCES

It's long been known that cannabis can make you sleepy. Cannabis tincture was sold as a sleep aid in the 1800s. Sir John Russell Reynolds, the queen's physician, wrote to the *Lancet* in 1890 to defend the utility of cannabis tincture. He said it was particularly good for 'senile insomnia': 'I have found nothing comparable in utility to a moderate dose of Indian hemp – viz. one-quarter to one third of a grain of the extract given at bedtime.'[29]

As well as cannabis having sleep-inducing qualities, it's thought that the endocannabinoid system is linked to circadian rhythms; so taking cannabis may help establish a better rhythm. However, despite the fact that improving sleep is one of the main reasons people buy CBD on the high street, there is a shortage of quality sleep lab studies, on either CBD or THC.[30,31]

Cannabinoids have an advantage over benzodiazepines; they don't make respiratory sleep problems, namely sleep apnoea or snoring, worse. However, they can worsen sleep paralysis, and we know that one of the main symptoms of THC withdrawal is sleep disturbance. A recent paper also showed that heavy recreational users of cannabis did not get better sleep – but that believing cannabis will give you better sleep does give better sleep.[32]

If you want to try cannabis for sleep, start with CBD oil. Sleep is one of the conditions being investigated in the Twenty21 studies; find a clinic at drugscience.org.uk.

ANXIETY

Most of the anti-anxiety effects of cannabis are probably down to CBD. What is interesting is that studies suggest it may work both in the short term and longer term.

In a preliminary study, people with generalised social anxiety disorder were given a single dose of CBD before a simulated public speaking test. Treatment with CBD reduced their levels of anxiety, cognitive impairment and discomfort to those of healthy control subjects.[33]

However, it seems you can have too much CBD; in another study, there were benefits at 300mg but these had disappeared at a dosage of 600mg.[34]

Anxiety is a major reason that people take illicit drugs. A study by the National Institute of Drug Abuse looked at CBD's effect on craving and anxiety in people who'd come off heroin. These symptoms often contribute to relapse. People were given high doses of CBD once daily for three days and results showed this reduced both anxiety and cravings, even three days later.[35]

CBD may have a calming effect on the central nervous system when taken every day too. One study looked at people diagnosed with anxiety and sleep issues, as the two often go together. Most were given 25mg a day; in the morning if their main condition was anxiety, in the evening if it was sleep. Over a month, 79 per cent saw a decrease in anxiety, and 67 per cent an impact on sleep, although the effects fluctuated.[36]

POST-TRAUMATIC STRESS DISORDER

Post-traumatic stress disorder (PTSD) can include hyperarousal, anxiety, depression and sleep disturbances, especially nightmares, and many people use cannabis to manage these. The data is pointing towards it being helpful for anxiety and possibly sleep (see above).

It also seems to help with other PTSD symptoms such as flashbacks, although evidence is limited.[37]

The relief it gives may be temporary, rather than a cure. An analysis of people who self-diagnosed with PTSD using a symptom monitoring app showed that immediately after taking it, people had a more than 50 per cent reduction in intrusive thoughts, flashbacks, irritability, and/or anxiety.[38]

In a study conducted in prison, the prescribing of nabilone, a synthetic THC, helped prisoners with serious mental illness. It reduced PTSD-associated insomnia, nightmares and chronic pain as well as other PTSD symptoms. It also allowed prisoners to stop taking other medications such as antipsychotics and sedatives/hypnotics.[39]

It may be that CBD works to help with the rise in anxiety and PTSD in healthcare professionals that we'll see post-Covid.[40]

A clinical trial testing this is under way. It will monitor the effect of 300mg of CBD, taken daily for 28 days, on the level of stress, depression, burnout and PTSD of a group of front-line health workers (physicians, nurses and physiotherapists).[41]

ADHD

The usual medication for ADHD is amphetamines such as Ritalin. So it may seem counterintuitive that cannabis, which appears to have an opposite effect, might help. But for some, it appears cannabis does work to regulate their behaviour and mental state.

It's well established both that people with ADHD self-medicate with street cannabis and that ADHD is a risk factor for problematic cannabis use.

In an analysis of online discussions of the effect of cannabis on ADHD, the results showed that there is no universal agreement: 25 per cent of commenters found it therapeutic, 8 per cent found it harmful, 5 per cent said it was both and 2 per cent said it had no effect.[42]

Professor Philip Asherson, one of the experts on Twenty21, trialled Sativex as a treatment for adults with ADHD. The result was that levels of hyperactivity/impulsivity improved and people were less disabled by their inattention. So we are now adding ADHD to the list of conditions covered by Twenty21.[43]

SCHIZOPHRENIA

We know that a lot of people with schizophrenia self-medicate with cannabis. We also know that high-THC cannabis can worsen psychosis and even provoke psychotic symptoms (see Chapter Thirteen). However, a more balanced mixture of THC/CBD doesn't seem to have this effect.

This is likely due to the protective effect of CBD: research in both animals and people indicates it has antipsychotic properties.[44]

In a brain imaging study, CBD was shown to have opposite effects on brain function to THC when given to healthy subjects. Pre-treatment with CBD prevented a dose of THC from causing psychotic symptoms.[45]

CBD slows the breakdown of anandamide, the body's natural endocannabinoid, and we know that raising levels of anandamide is associated with reduced psychotic symptoms. Brain imaging studies also show CBD may work by rectifying some of the brain disconnection issues of schizophrenia.[46]

As a result, CBD is now being tested as an antipsychotic treatment for schizophrenia, with some positive results. In a double blind trial, people with schizophrenia were given high doses of CBD or a placebo; after six weeks the CBD group had lower levels of psychotic symptoms.[47]

HEADACHE AND MIGRAINE

There's evidence that cannabis has been used for migraines for thousands of years, both to prevent and reduce migraines and also to treat them.[48]

Modern studies, though limited, seem to back this up. One study looked at people suffering from migraine who'd been taking medical cannabis for three years. More than 60 per cent had fewer migraines, and also had reduced their intake of other painkillers.[49]

More recently in the US a research group used an app to gain more fine-grained analysis of the usefulness of cannabis flower in treating headache and migraine. Ninety-four per cent of people said they had a reduction in pain within two hours of using cannabis. And a higher THC level – over 10 per cent – was the strongest predictor of symptom relief.[50]

Another app study showed that inhaling cannabis reduced severity by approximately 50 per cent. However, it also showed that people developed tolerance over time.[51]

To prevent this happening, it may be better to use cannabis only to treat acute pain rather than taking it regularly.

DEPRESSION

A review of studies of 76,000 cannabis users showed they do have an increased risk of depression, although this effect only became significant with heavy use.[52]

Another large review, this time of studies looking at teenagers, showed a higher risk of depression but not of anxiety.[53]

Despite the fact that being stoned seems to increase the risk of depression, a UK survey showed people commonly use cannabis to self-medicate for this.[54]

So why do people think cannabis helps their depression? There is a plausible biological basis for cannabis affecting mood. It's been shown that the drug rimonabant, which works by blocking endocannabinoids, leads to low mood. THC intoxication does lighten mood and make people somewhat euphoric. And it improves sleep and anxiety, both of which are impaired in depression.

But the key to positive changes, as in other psychological conditions, appears to be CBD. As well as having antipsychotic and anti-anxiety properties, it may also be antidepressant.[55]

One possible mechanism is that CBD helps reduce brain inflammation. Eminent psychiatrist Professor Edward Bullmore explains in his book *The Inflamed Mind* how some depression can be caused by bodily inflammation, so that inflammatory substances cross the blood–brain barrier and cause mood and behavioural changes. Bodily inflammation can be caused by physical and social stress as well as lifestyle, obesity and inflammatory health conditions.[56]

As yet, there are no good trials on medical cannabis that show exactly what it does to mood. We are including ratings of mood changes in all of our Twenty21 trials, so we hope to have data on this soon.

EHLERS–DANLOS SYNDROME (EDS)

In these rare, hereditary connective tissue disorders, people don't produce collagen properly. The result can be hypermobile joints that rupture easily, as well as digestive issues, fragile and sensitive skin and severe pain. People may need their hips or shoulders replaced, or their jaw might dislocate so they can't swallow. They can often be on huge doses of opioids for the pain.

As you can see from Lucy Stafford's story later on in the chapter, cannabis treated her pain effectively and so improved her life enormously. Based on Lucy and the other patients I have seen who have found major benefit from medical cannabis, it would seem to me that all patients with this syndrome should be given a trial of medical cannabis if they want one.

CANCER

As described above, nabilone can be prescribed for cancer- and chemotherapy-related nausea and loss of appetite. And there's good evidence that cannabis works for cancer-related pain too. This makes cannabis extremely useful as an all-rounder: 60 per cent of people with cancer reported an improvement in symptoms including pain, nausea and lack of appetite, as well as weakness and sleep problems.[57]

But there is a bigger and more controversial issue: whether cannabis can treat cancer. I have come across many people whose cancer has seemingly responded to medical cannabis. Some are now fully cured. This is not implausible, although research is still limited by its illegal status. There is some biological basis; animal and pre-clinical research shows that THC, CBD and other cannabinoids may kill cancer cells and stop cell proliferation.[58]

Other studies have shown effects in leukaemia, lymphoma, glioblastoma and cancers of the breast, bowel, pancreas, cervix and prostate.[59]

But we don't know a lot about how cannabis works against cancer, whether it will work at all or the dosage, though in general people with cancer tend to use higher dosages.

The question is: what should you do if you get cancer? Many people write to me saying they have cancer and would prefer cannabis over conventional treatments. My advice to them is usually: have the conventional treatment and take cannabis as well.

Given the safety and tolerability of medical cannabis and the fact that it is unlikely to interact with the other treatments for cancer, and the fact you may well be prescribed nabilone for nausea, it seems to me acceptable that people should try it. However, you should share with your oncologist the fact that you're doing this.

HIGH BLOOD PRESSURE

People report taking cannabis for high blood pressure. This seems counterintuitive, as cannabis can increase heart rate. And indeed case reports have also linked it to heart attack, heart arrythmia, cardiomyopathies, stroke and inflamed arteries.[60]

However, these were mainly non-medical users, likely using high-THC products rather than pharmaceutical-grade cannabis. Again, new research shows CBD may moderate the impact of THC and so reduce negative effects on the heart.[61] If you have heart problems, take cannabis with care, and ideally low-THC products.

SPINAL CORD INJURY

Cannabis may be useful for pain, spasticity and sleep disturbances associated with this type of injury (see separate sections above). People with any kind of neurological disorder often get bladder spasm, and cannabis seems to relieve that too.[62]

HUNTINGTON'S DISEASE

This is a genetic condition that causes the degeneration of nerve cells in the brain. Results of studies in the lab and animals looked encouraging, but these haven't translated into human results. Cannabis may be helpful for pain and spasticity associated with this disorder, as in the sections above.[63]

For this reason, the FDA in the US has recently given orphan drug status to a medical cannabis product currently in research for Huntington's. This means that only one positive trial is required in order for the drug to be licensed.[64]

MOTOR NEURONE DISEASE

This is a progressive paralysing disease that damages the motor nerves so that eventually people can't move or even breathe. Unpleasant as this is, it's made worse by the fact that the person's brain is completely unaffected so they are fully aware of their progressive decline and inevitable death from breathing complications. There is no treatment, and patients usually die within a few years, but can be kept alive longer on artificial respiration. As with Parkinson's disease, cannabis has been shown to give relief from symptoms such as pain and spasticity.[65]

Also, animal models have suggested cannabis may slow its progression.[66]

OBESITY AND DIABETES

The munchies is a real phenomenon, so why aren't regular cannabis users overweight? In fact, several epidemiology studies show they tend to have lower body mass indexes than the general population.[67, 68]

It could be because cannabis has no calories (unless it comes in the form of a brownie) and people are using it instead of drinking alcohol, which is high in calories. However, it's likely there's another effect going on here, too.

The link between the endocannabinoid system, appetite and weight has been known for a while, but the exact mechanism isn't known. The endocannabinoid blocker rimonabant was licensed as an anti-obesity drug but withdrawn in 2008.

Cannabis may have a role in preventing obesity-related (type 2) diabetes. In a study of the Inuit population, regular cannabis smokers had lower BMI and lower fasting insulin. Their lower weight led to less insulin resistance – the precursor to type 2 diabetes.[69]

There is also the possibility that CBD could help in type 1 diabetes – at least in a rat model. A 2020 paper suggests cannabis may have direct effects on insulin resistance here. Two weeks of treatment with CBD improved the condition of rats that had been made acutely diabetic.[70]

CAN CANNABIS HELP PEOPLE WITH LONG COVID?

There's a whole range of neurological and psychological consequences that can persist after a Covid infection. These include: fatigue, anxiety, cognitive impairment (brain fog), pain, which can be all over the body, headaches, and breathing problems.

Many of these symptoms occur in other conditions too and have been shown to be responsive to CBD.

Drug Science is conducting a formal trial on this, in collaboration with an Australian medical cannabis manufacturer, Bod.

We plan to recruit up to 40 patients with long Covid and treat them with CBD for at least six months, to see if there is any evidence that it helps. Results should be ready by the middle of 2022.[71]

I owe my life to medical cannabis

Lucy Stafford, 21, has been medicating using cannabis for two and a half years. She has the connective tissue disorder called Ehlers–Danlos syndrome (EDS) (see page 146).

My condition affects the production of collagen all over my body. It means my blood vessels, skin and connective tissue are too lax. Internally, my digestive tract and bladder function are also affected. I can dislocate my shoulder by brushing my hair, my jaw by yawning or my knee by rolling over in bed. Dislocations cause muscle spasms that are excruciatingly painful.

CANNABIS

I had my first surgery aged 10, then 19 surgeries through-out my teenage years. EDS symptoms tend to get worse from puberty, as hormones affect hypermobility. I used a wheelchair from the age of 16, and I was bed-bound from 17. Often the pain was so bad I would cry myself to sleep. It ruled my whole life.

The only medication prescribed for pain as severe as that is opiates, which I took daily from the age of 13. It was almost impossible to engage in the world around me while taking opiates, I felt so numb, so I dropped out of school when I was 16. By the age of 17 I had been prescribed every opiate there was for the pain and benzodiazepine for the spasms. Even on fentanyl, the pain was severe. The opiates made me so exhausted I couldn't exercise or even get out of bed, so my muscles were wasting away too.

The combination of gastroparesis (a paralysed stom-ach), which is a co-morbidity of EDS, and opiates was disastrous for my digestion, and I stopped being able to digest food or water. I had a central line placed to receive intravenous nutrition. But that first year, complications caused blood clots and I had sepsis six times – I almost lost my life in intensive care.

Then when I was 18 my jaw dislocated, just from yawning. It was agonisingly painful and could not be relocated, so I was admitted to hospital. Every day, the doctors came in to try to manipulate the joint back into place, but the muscle spasm was too severe. I was on intravenous morphine, zonked out but still crying in pain. I had an operation to wire my jaw shut but, due to the nature of my condition, within a week of having the wire removed it had dislocated again.

WILL CANNABIS WORK FOR ME?

One day in an emotional appointment with my pain specialist, he said he could not increase my opiate dose any more. The risks outweighed the benefits. Opioids become less effective over time, and my pain had got worse and worse, so we didn't know what to do.

This was at the end of 2018, when medical cannabis became legal to prescribe in the UK. My doctor had read it was an antispasmodic, so suggested it as an option. He wrote the prescription but we were turned down for NHS funding. In fact, I got a letter saying cannabis was unlikely to work and, if I took it, there was a one in four chance that I'd end up with psychosis.

But by that time, the seed was planted in my mind that cannabis might help. Really, it was my last hope. My mum and I travelled to Amsterdam, which was not an easy trip to make in a wheelchair and with a feeding tube, but we were desperate.

At a coffee shop, I tried cannabis for the first time using a Volcano vaporiser. Immediately, I felt my spasms begin to release, my pain ease and my nausea subside. For the first time that I could remember, it felt like my body was beginning to regulate itself.

When we got back from Amsterdam I carried on using cannabis. But trying to source it, I was scammed and put in vulnerable situations, and the supply was never very consistent. Eventually I found a source of good-quality cannabis, although it was still illegal.

Taking it was game-changing for my symptoms. Slowly but surely, the spasms in my jaw began to unlock. That was all I ever hoped for – some relief from my jaw pain.

I started to be able to reduce my opiates and other pharmaceutical medications. Once I stopped taking them, I found I no longer needed anti-sickness medication. And that I was no longer exhausted and sleeping all day. I came out of the opiate haze. I started to feel my emotions. Now I could focus, I started an Open University degree in Science, Technology, Engineering and Maths.

All the things I'd been told I wouldn't be able to do – eat, drink, wee, walk, exercise – I found I could slowly start to do again. I could strengthen my muscles, which then reduced the number of dislocations.

It wasn't until I got a private prescription for cannabis that I felt I could be open with my doctors about what I was taking. Before, I'd been too scared to tell them due to the stigma. Doctors are educated in the risks, not the benefits, of cannabis.

I'm now prescribed three cannabis medicines; a CBD oil and a THC oil I take three times a day, and I vaporise flower for acute pain relief. If I am in severe pain, vaping can decrease it from 9/10 to 4/10 in ten minutes, whereas before I'd need intravenous opiates to do that.

My private prescription for cannabis initially cost £1,450 a month, which has gone down to £450 now I've joined Twenty21. Thankfully my family are supporting me to fund my medication, but I'm a disabled student so I know paying this much is not sustainable.

That's why a team of patient advocates and I launched non-profit advocacy organisation Patient-Led Engagement for Access (PLEA) to campaign for NHS prescriptions for everyone who needs them (pleacommunity.org.uk).

My cannabis prescription would cost the NHS significantly less than my previous treatment costs. When I was on a feeding tube, my medication alone cost over £250 a day, and that doesn't even include the hospital stays. Since I've been able to treat my pain at home, I haven't had to go to hospital once. It feels incredible after having to grow up in them.

For the first time in my life, I can plan for the future. I can think clearly and engage with the world. I'm about to start a Medical Neuroscience degree at the University of Sussex, and in the future, I hope to specialise in neuropsychopharmacology.

I've moved to Brighton and I'm living independently for the first time. I've learned to roller-skate and I skate along the seafront. I go swimming in the sea. I walk up to seven kilometres a day. Because I can eat, I'm now a healthy weight. Before cannabis, my quality of life and prognosis was poor. Cannabis isn't a miracle cure as I still have my underlying condition – but now I'm able to live my best possible life.

SECTION FOUR

HOW TO MINIMISE THE HARMS

12

WILL I BECOME DEPENDENT?

CANNABIS MIGHT BE one of the world's oldest medicines, but it's a relatively new addiction. After qualifying in medicine in 1975, I worked for a few years as a junior doctor in A&E. I cannot remember a single person presenting whose condition was anything to do with cannabis – although I can remember thousands whose issues were caused by being drunk and/or addicted to alcohol.

Two years later, I trained and then worked as a psychiatrist. At that time, cannabis rarely came up as a problem on its own. When we did a full introductory psychiatric interview, we'd ask the patient if they drank alcohol, and if they smoked cannabis, as well as whether they took other drugs. Occasionally, I'd see people who had become psychotic who said they were a regular smoker of cannabis. But dependence on cannabis was practically unheard of.

Now, dependence is a very real problem, globally. This isn't surprising: cannabis is the most used psychoactive substance in the world, according to the World Health Organization (alcohol is used way more but it isn't treated as a drug by the WHO).[1]

That makes cannabis by far the most common 'illegal' drug, so it's the one most people are likely to encounter and so risk becoming addicted to. Cannabis dependence can be as problematic as alcoholism. In the UK, cannabis is the leading drug for which younger people, those under the age of 30, seek help.

However, let's put those numbers into perspective. In 2019 to 2020, 132,124 people started treatment for drug and alcohol problems.

Of the people starting treatment:

- 59 per cent said they had a problem with alcohol
- 32 per cent said they had a problem with opiates
- 22 per cent said they had a problem with crack cocaine
- 20 per cent said they had a problem with cannabis
- 16 per cent said they had a problem with cocaine

Last year 15,000 people in the UK were in treatment for cannabis dependence compared with 140,000 for opiates and crack and 75,000 for alcohol.[2]

In this chapter, I want to lay out the real risk of cannabis dependence, and explain how and why it's become a problem more recently. And how the risk varies according to the type of cannabis you're using and also why you are using it. If you are thinking of trying medical cannabis, I hope this will give you a scientific basis from which to decide.

Currently, the perceived risk of becoming dependent stops people trying cannabis to treat medical conditions. Along with the fact that cannabis has been vilified since the 1970s, the fact that it's illegal in the UK and that you have to buy it on the black market, is an added hurdle to what may turn out to be an effective treatment for their condition.

WHAT DOES CANNABIS DEPENDENCE LOOK LIKE?

I describe dependence (aka addiction) like this: a behavioural disorder underpinned by changes in the brain, which leads to continued use of a drug or substance in the face of problems such as withdrawal. Added to that, your use of that substance interferes with your family and social life and causes you personal harms.

Typically we talk about dependence in terms of physical and psychological effects. Physical dependence is the adaptive changes that occur in the body – in the case of cannabis, in the brain – that do two things.

The first is, the effects of the drug reduce over time, which is called tolerance. The second is, withdrawal. That is, when you stop the drug, the brain goes into withdrawal and you get symptoms.

Markers of psychological addiction are the same as for other substances. You spend more and more of your time preoccupied by it, trying to buy it or using it, you spend more of your money on it. You start to give up other activities in favour of getting stoned. You may lose your job or your family relationships may be affected by it. If you are unlucky, you may get a criminal record for possessing cannabis.

People who begin using as teenagers are much more likely to develop problem use. In fact, they're four to seven times more likely to become dependent than those who start using as an adult, perhaps because their brains are still developing.[3]

However, this is true for all drugs. So it may also be down to the fact that drugs are commonly tried by young people, so if you have a propensity to dependence, this is the time in your life that you'd try it.

There is a typical cannabis addict. If you are reading this book because you are worried about a young person, friend or loved one, they may sound familiar to you.

He's a teenager or young adult, most often male – as men are more likely to become users of all drugs, although women are catching up. He probably started using the drug with friends. He moved on to using it alone, and to using it every day. He might have started smoking in the mornings before school, college or work, then going out of work or college at lunchtime to smoke. At this point, he might realise he's dependent, that cannabis is stunting his life.

This progression mirrors the story of addiction to alcohol in that it's like a ladder, so the person gradually moves a rung at a time towards dependence. There are differences: while alcoholics tend to be in their thirties or forties when they go into treatment, cannabis addicts are often in their twenties.

One reason might be that drinking is more socially acceptable than cannabis addiction. So some people who drink heavily manage to hold down high-level jobs for many years; it's not clear if this is true for cannabis. And it's a lot easier to hide alcohol addiction in plain sight. You can be the person who drinks a lot at dinners, parties and work socials and be seen as 'good value'. Being a 'stoner' is more difficult to hide, and still has a stigma attached.

The effects on the people around a cannabis user and an alcoholic look different too; cannabis doesn't make people violent in the same way as alcohol. But heavy use of cannabis can lead to apathy, job loss or leaving education.

There are many more people who aren't necessarily dependent but for whom cannabis use is a problem. They might use more than they'd like, have a spliff every night after work or smoke during the day every weekend.

There are even more people, in fact the vast majority, for whom cannabis doesn't appear to be a problem. They don't become addicted and they don't smoke every day. For a lot of people, use tails off as they get older and gain responsibilities such as jobs and families.[4]

THE FIRST PART OF DEPENDENCE: TOLERANCE

In simple terms, becoming more tolerant means you need bigger and bigger hits to get the effect you seek. Tolerance exists for cannabis but the effect is much weaker than, for example, for heroin.

Tolerance is likely due to a reduction in the number of cannabis receptors, although we can't yet be certain of the mechanism.[5]

Cannabis only partially stimulates your body's endocan-nabinoid receptors. Heroin, on the other hand, plugs perfectly into the body's opioid receptors.

Clinically, we see that many people taking cannabis for medical reasons do have to increase their dosage over time. Some patients taking part in Project Twenty21 have told us this. But this isn't the case for everyone and, for medical users, tolerance isn't the same as dependence.

When it comes to recreational use, as people become more experienced at taking cannabis, they tend to put more canna-bis into each joint, inhale more deeply, and take more puffs. It may be that tolerance develops if you take a lot. One study gave moderate and heavy users access to as many standard-ised joints as they wanted. It was only the heavy users who reported feeling less intoxicated over the course of a month.[6]

However, even tolerant users likely moderate their intake. During reviews of evidence for the ACMD in 2005, some Dutch experts came over to the UK to present to the commit-tee. We were interested in finding out more about the high-THC, low- or no-CBD version of grass that was taking over the UK market.

The Dutch told us it wasn't inevitable that people got more stoned on high-THC cannabis. They described giving people either lower- or high-THC forms and seeing that once the smoker had tried the high-THC version, they ended up inhal-ing less of it. So, to some extent, it appears people self-titrate: they judge the effect, then moderate their intake to arrive at the level of intoxication they like.

THE SECOND PART OF DEPENDENCE: WITHDRAWAL

Cannabis withdrawal is certainly nothing like as dramatic as that from opioids, where you can ache, become anxious, sweat, vomit, and have the shakes and diarrhoea. Or alcohol withdrawal, which can cause seizures, delirium and even death.

That's because, firstly, cannabis isn't as toxic to the brain as those other drugs, so the brain doesn't adapt as much to compensate; and secondly because cannabis clears relatively slowly from the brain.

That said, if you use regularly and particularly if you use heavily, adaptive changes can occur that lead to both increased tolerance and a withdrawal syndrome that lasts up to two weeks.

These adaptive changes have been shown in experiments where a subject is given an antagonist to block the cannabis receptors, which then rapidly brings on withdrawal symptoms. People become irritable, have mood swings, find it hard to concentrate, get headaches, chills and sweats, and can't sleep. They also get cravings for cannabis.[7]

WHAT DRIVES ADDICTION?

The sheer number of people who take cannabis isn't the driver of the current levels of dependence. There's a deeper reason to why I didn't see cannabis dependence in the 1980s but it's relatively common now; the true driver is the nature of skunk, because it's so much more addictive than lower-THC forms of cannabis.

In the 1990s, skunk grew to become the dominant product on the black market (see Chapter Thirteen). And as you can see from the figure below, as THC levels rose, dependence did too.

Source: https://pubmed.ncbi.nlm.nih.gov/29382407/

Fig. 10 Time course of changes in cannabis potency and referrals for cannabis dependents in the Netherlands

This figure, from a study co-authored by Dr Tom Freeman of the University of Bath, shows how the rise in THC led to a rise in people seeking help for addiction. The solid black line shows the THC content of randomised samples from coffee shops in the Netherlands, from 2000 to 2015. The second line shows the number of people in the Netherlands who presented to addiction services during the same time period. You'll see that the second line mirrors the first, with a roughly two-year gap.

Cannabis dependence didn't really become an issue until cannabis became a political issue. Increasing dependence during the 1990s was the consequence of cracking down on what used to be a relatively harmless drug, compared to alcohol and tobacco. This created a more dangerous form of that drug, which in turn led to the creation of real harms.

If skunk hadn't started to dominate the market, addiction probably wouldn't be the issue it is today, and neither would there have been the rise in psychotic episodes due to cannabis, described in Chapter 13.[8]

On an individual level, the reasons that people start using, keep using and become addicted is a grey area, as we don't have a lot of data due to cannabis's illegal status. But we do know that why someone becomes dependent will be related to why they started in the first place.

For everyone who starts using it, up to 10 per cent will become dependent, the likelihood being higher if they use skunk. Each person's reasons are likely to be a combination, similar to every other addictive substance, of genetics, personal history, personality and lifestyle.

But we know almost nothing about the genetics of cannabis-liking and we need more research on the brain effects too. We don't have nearly as much data on this as we do for alcohol. We know it's an exploratory drug for young people. And we know that everyone will fall somewhere on a spectrum between recreational use and self-medication for physical and/or mental reasons.

It's relatively common for people to use cannabis to self-medicate for mental distress. As mentioned already, Dr Daniel Couch, medical lead at the Centre for Medicinal Cannabis, led a YouGov survey in 2019 to dig into this question, surveying people about their use of street cannabis to treat a diagnosed mental condition. Depression and anxiety came up as the two most common mental health reasons, with 8 per cent of the respondents using cannabis for depression, and the same proportion for anxiety. Other mental health conditions were insomnia, PTSD and schizophrenia.[9]

There's also a group who come to cannabis in later life: people who've been traumatised. These patients say that cannabis helps with symptoms of PTSD and is good for nightmares; in the CMC survey, 17 per cent of people surveyed were using it for this.

HOW ADDICTIVE IS CANNABIS?

I've said that around 10 per cent of people who try cannabis become addicted. Despite the fact that skunk is so much more addictive than cannabis, you may be surprised to know that this risk of dependence is still lower than that of other drugs, including alcohol and nicotine.

If you measured the addictiveness of a drug on a scale, with heroin at 100, alcohol comes in at around half that, at 50, and cannabis comes in at around a third.[10]

Put another way, 30 per cent of people who try heroin become dependent and 15 per cent of people who try alcohol do so, compared to around 10 per cent who try cannabis. And, as we have seen, the most addictive drug of all is a legal one: cigarettes. Up to 60 per cent of people who try smoking go on to become smokers.[11, 12]

There's one caveat: these numbers are confounded by the fact that alcohol and cigarettes are so easily available. So the question is, if there was a sensibly regulated market that made cannabis more available, would levels of dependence rise?

This is not necessarily the case. Firstly, it's teenagers who are most at risk of developing dependence. While the illicit market doesn't care about the age of its consumers, a regulated market can impose a lower age limit for purchase. Secondly, in a regulated market such as the coffee shops in the Netherlands, people have the choice to buy less addictive, lower-THC forms of cannabis. In fact, in Uruguay, THC is capped at 15 per cent.

The idea of cutting risk by discouraging use of the strongest form of a drug isn't a new one. There was some hysteria around high-strength alcohol leading people to ruin during the gin craze of eighteenth-century London, as seen in Hogarth's depictions of Beer Street versus Gin Lane. To solve what was seen as an epidemic of drunkenness from gin, the government passed five acts between 1729 and 1751. These included increasing taxation on spirits to regulate and reduce their consumption. Since that time, the UK government has always taxed strong alcohol more than weak in an attempt to control excessive use and resulting harms.

However, when it comes to cannabis there is an extra reason in favour of limiting the strong version. People who drink spirits are not more likely to become addicted than people who drink weaker alcohol (though those who are dependent may drink more spirits as their tolerance increases). But people who smoke high-THC skunk are more likely to become addicted.

IS THE SOLUTION WITHIN?

The reason that high-THC cannabis is more addictive isn't just the THC content. As you up the amount of THC during the growing process, the CBD content plummets. And evidence is growing that CBD in the traditional forms of cannabis is protective against addiction.

There is currently no medication available to be prescribed to help people reduce or stop their dependence on cannabis. But the research into CBD's effect on addiction looks interesting.

The UK study led by Dr Tom Freeman tested the idea that giving CBD could help reduce use of cannabis in people with dependence. During the four-week study, those given daily doses of 400mg or 800mg of CBD (which is very high – a typical retail dose is around 25mg CBD) along with counselling sessions did increase the amount of time they were abstinent.[13]

It's thought that as well as working as a sort of antidote to the addictive effect of THC, CBD may possibly treat aspects of the withdrawal that happens when THC use is reduced.[14]

An Australian study used Sativex, which contains an equal dose of THC and CBD, to help people reduce their use of high-THC street cannabis. You could compare this strategy, of prescribing a lower-harm version of a drug, to vaping instead of smoking tobacco.[15]

HOW DO I KNOW IF MY CANNABIS USE IS A PROBLEM?

The CUDIT-R questionnaire (see Appendix) will give you an indication of whether you're dependent or have a problem with cannabis. If you answer yes to any of the questions, it's not proof of this but it does give some indication. If you think you have a problem, your GP is a good person to talk to about it. But – and this is a big but – the questionnaire is aimed at those using cannabis for non-medical (recreational) reasons. It has limited utility for medical cannabis users, which I'll discuss below.

My charity Drug Science is now exploring the question of how to assess dependence for medical users, using an updated version of CUDIT modified for medical cannabis users by Professor Val Curran from University College London. This is important to counter the pervasive belief that chronic use is the same as dependence, a key barrier to using cannabis as a medicine.

That's because medical users will score highly on some of the original dependence questions as a matter of course: namely, using cannabis every day and not being able to stop once started.

Using every day and even an increased dose may not be a sign of tolerance, but simply necessary to treat symptoms. And a medical user will also be able to engage a lot better in work and education as well as in a social life and relationships. But perhaps most important of all, the medical user isn't using cannabis to get high but to alleviate symptoms.

Medical Use	Recreational Use
Often daily	Full range from rarely to daily; only some daily users are addicted
Various routes of administration	Often smoked with tobacco in the UK
Main aim to alleviate symptoms	Main aim enjoyment, relaxation, social effects
Little desire to get 'high'	Liking effects of THC

Lower prevalence of substance use disorder	Often other psychoactive drug use
Lower controlled dose of known quantity of cannabinoids/cannabis	Higher THC dose, often unknown, cannabinoid content usually unknown
Regulated quality	Unknown quality
Poorer general physical health	Usually good general health
Poorer psychological well-being	Generally good psychological well-being
Lower physical quality of life	Generally normal range of quality of life
Higher age >50	Lower age

WHAT HAPPENS IF YOU NEED TREATMENT?

Your first stop is your GP, who can work out if you have a problem.

You can help them by noting down how much you are using, and when you are using it.

Also ask yourself why you're using cannabis.

If it's to deal with a psychological or medical problem, then your GP can offer alternatives. These might be counselling, or other forms of medicine.

There are no drugs used currently to treat dependence, and, considering the wide use of cannabis, there is a limited amount of research. This is probably because those dependent on cannabis don't tend to cause others harm. Conversely, there's a lot of research into opiate dependence, probably because the social impact is much greater.[16]

If your problem is that you've been using a lot of cannabis and you can't stop because of cravings or withdrawal, you may need help from someone with expertise in addictions.

Your GP will therefore refer you to your local drug service. In the meantime you could also try an organisation such as Marijuana Anonymous.[17]

Your local drug service will do a more detailed assessment of the reasons you use cannabis, what you've been using and for how long, and they will explore with you what happens when you stop using it.

They can help you get through withdrawal, possibly using medication but more likely through counselling and motivational support.

They may also discover that you do have underlying psychiatric problems for which you were self-medicating; they can provide you with more expert help for these.

You will also be helped to develop strategies so you don't slip back into using, ideally with motivational enhancement therapy (MET) or cognitive behavioural therapy (CBT).[18]

Both will help you tackle the mistaken assumptions you make about cannabis use, such as: 'I need it.' Or: 'If I don't take it I can't function with other people.'

At the drug service, they will also help you deal with any of the problems that occur as a result of your use, such as falling out with your family or friends, or problems at work, and any legal issues that may have arisen.

THE GATEWAY THEORY: DOES CANNABIS LEAD TO OTHER DRUGS?

One argument regularly wheeled out by politicians to defend keeping cannabis's illegal status is that cannabis is a gateway drug, the 'soft' drug at the beginning of a slippery slope to harder ones.

It's true that where cannabis is illegal – that is, in many countries – it's usually the first illegal drug people try. However, both tobacco and alcohol are usually used even before cannabis, and at an age when their purchase is illegal. And we know that young people who use any drugs – legal and illegal – are different to young people who don't; more adventurous and sensation-seeking, more vulnerable.

The argument that says cannabis is a gateway drug goes like this: every heroin user has smoked cannabis and therefore cannabis is a gateway drug.

This is a logical fallacy. Firstly, every heroin user has likely also smoked tobacco and drunk alcohol. But although these are just as much 'gateway drugs' in that heroin addicts will almost certainly have used them, they're almost never called that.

The real reason there's an association between cannabis and heroin is down to the illegality of cannabis. To buy cannabis in the UK, you have to get it from a dealer. And most dealers who sell skunk also deal heroin and crack.

This is why the Dutch developed their coffee shop approach. They wanted to split cannabis off from hard drugs like heroin and cocaine. And it worked: the Dutch have a lower use of heroin in young people than comparable countries. Additionally, when they set up the coffee shops, there was no increase in cannabis use.[19]

CANNABIS AS A TREATMENT FOR DEPENDENCE

Many people who start using medical cannabis for pain are able to wean themselves off opiate painkillers (see Lucy Stafford's story in Chapter Eleven).

In one study, 97 per cent of the sample were able to decrease the amount of opiates they were taking.[20]

As people switch to cannabis, they slash their risk of addiction to opiates, as well as of overdose and death.

To reiterate, the amount of cannabis it takes to kill you is over 1,000 times an average dose; in comparison, the amount of opiates or alcohol that it takes to kill you is just two to four times an average dose.[21]

Together, this shows what the gateway theory actually is: a concept invented by prohibitionists to justify keeping cannabis illegal. In fact, it's prohibition that's driving rates of dependence, as that's what led to skunk dominating the market.

The lesson for the individual user is: avoid skunk. A regulated market could include a cap on THC levels, as in Uruguay, or could price low-THC products to encourage their use over high-THC ones.

It could also ensure lower levels of cannabis use in young people by imposing an age limit, and so reduce likelihood of harms including dependence.

At the very least, making medical cannabis available to people who need it, rather than leaving them at the mercy of illicit markets, would ensure they were taking safer forms of cannabis, under the guidance of a doctor, and so would reduce their risk of dependence.

ARE YOU READY TO GIVE UP CANNABIS?

These are some very quick questions to help you to think about your behaviour around cannabis. Yes answers show that smoking is affecting your life in a negative way.

1. Does smoking cannabis feel like something you have to do?
2. Do you mainly smoke on your own?
3. Are most of your friends people you smoke with?
4. When you smoke, do you zone out into your own world?
5. Can you imagine your life if you didn't smoke?
6. Do you smoke then not do the things you intend to do? Such as work, studying, going out?
7. Does smoking make you forget things, or not be able to concentrate?
8. Do you plan your life around where and when you can smoke, or buy more cannabis?
9. Do you get worried if you're running out of cannabis?
10. Do you smoke to cope with or avoid your feelings or problems?
11. Has your partner, friends or family or anyone else said they're worried about your smoking?
12. Have you promised friends or family that you'll cut down or stop smoking, then not done it?

13

WHAT ARE THE REAL HARMS OF CANNABIS?

UNTIL 2019, WHENEVER the government were faced with evidence that their policy on cannabis was unscientific, their reply was always to repeat a version of this mantra: 'Drugs are harmful. Cannabis is a drug and so it is harmful. So we will keep cannabis illegal.'

The dogma of prohibitionism has been so pervasive in our society that for most people, it simply feels true. As a society, we have been brainwashed into believing that the 'war on drugs' is the right approach. 'Just say no', as the 1980s campaign told us.

As I wrote in Chapter Two, the stance of being tough on drugs has been used as a vote winner for both the major UK political parties. And this approach has been backed by the newspapers, who've for the most part been happy to print fearmongering headlines about drugs.

As a result, there are a whole range of scare stories about cannabis that have been made up and/or exaggerated to justify political decisions and to try to put people off using it.

This has prevented a lot of people even considering using cannabis medically. And the lingering stigma in the UK still stops many doctors from considering it as a possible treatment.

Cannabis isn't risk-free, but what medicine is? And it's a whole lot safer when a patient takes quality medical-grade products with the advice of a doctor rather than illicit products of unknown make-up bought from a street dealer.

In this chapter, I look at the current evidence behind the scare stories and anti-cannabis propaganda. It's only by taking account of the real downsides of cannabis that we can begin to use it more safely. Looking at the evidence will also show us the areas where research is most needed.

CANNABIS AND MENTAL HEALTH

When I talk to the public about cannabis in the UK, there's one question that always comes up. And it is: Is cannabis making our kids schizophrenic? The short answer is no. This focus on and fear of cannabis-induced psychosis is a peculiarly British one; it's not nearly such a preoccupation in other countries.

There are two reasons for this. Firstly, the idea that cannabis causes schizophrenia was supported by UK academics.

And secondly, it's a subject that's regularly trotted out by UK newspapers, in particular by the *Daily Mail,* which has had a major influence on the public debate about cannabis by violently opposing any policy change.

I've been having this argument about schizophrenia and cannabis throughout my career, most notably in 2009 when the government gave the risk of schizophrenia as one of the key reasons to move cannabis from Class C to Class B and I was sacked from my post as chair of the Advisory Council on the Misuse of Drugs (ACMD) for voicing my disagreement. The council had, by this time, conducted three full reports on cannabis, including a separate review. We had looked at thousands of pieces of evidence in order to decide our position. Yet, still the Home Secretary opposed our recommendation.[1, 2]

MENTAL HEALTH IN HISTORY

One reason why the mental health story about cannabis is so enduring may be because it has roots going back over a hundred years. In the 1890s the Indian Hemp Drugs Commission was the first official British inquiry into the effect of cannabis on mental health.

At the time, the British in India were getting rich by exploiting the raw materials of the colony. And that included cannabis, which Indian people had been using as a medicine and for religious reasons for thousands of years. From the 1700s, the British East India Company had been making money by taxing the growing, sale and use of all cannabis products, such as the milk drink bhang, ganja and charas.

CANNABIS

In the later 1800s a growing movement of Victorian puritans began agitating to ban all psychoactive substances, from alcohol and laudanum to opium and cannabis. In the early 1890s, after reports that asylums were seeing an influx of Indian people due to their cannabis use, puritan sympathisers brought this up in parliament.

By this time, Queen Victoria was Empress of India, and in theory responsible for the well-being of her Indian subjects. So the government set up the Commission to investigate the desirability of prohibiting cannabis. It was a serious parliamentary committee, made up of influential and powerful men.

The study was nothing if not comprehensive, looking into the effect of cannabis on digestion, constitution and intellect, and its connection to crime and violence. With hearings in 30 cities over a year, there were close to 1,200 witnesses questioned, and the report itself ran to seven volumes and more than 3,000 pages. Commissioners visited asylums and examined the cases of patients who'd allegedly been admitted for cannabis use.

Brigade Surgeon-Lieutenant-Colonel D. D. Cunningham, FRS, also carried out three experiments on monkeys. One involved examining the brain of a rhesus monkey that had been given cannabis over eight months. The autopsy revealed a normal-looking brain.

This is also what is found with human beings. Unlike with alcohol use, post-mortem studies of the brain after cannabis use do not find abnormalities.

Our group uses MRI scanning to look at the brain abnormalities that relate to addiction, in order to develop new treatments for alcohol, cocaine and heroin.

When we started one trial looking at the brains of alcoholics, we found that a large proportion of the people we were studying were also using cannabis.[3]

This presented a challenge. If we excluded them from the research we'd have too few subjects.

We decided to include them, but compare their results with those who didn't use cannabis.

At the start of the project, I made a bet with my co-investigators that the alcoholics who also used cannabis would have less pronounced brain abnormalities than those who didn't use it.

And I won my bet. This is consistent with the data that's now emerging that cannabis is neuroprotective in conditions of oxidative stress such as epilepsy,[4] and also protective against liver cirrhosis in alcoholics.[5]

Going back to the Commission, their overall conclusion was that there was no harm when cannabis is used in moderate quantities, either on the mind or the body, or on society.

'In regard to the physical effects, the Commission have come to the conclusion that the moderate use of hemp drugs is practically attended by no evils at all.'[6]

Out of the three forms of cannabis, bhang got the most positive write-up. The report stated: 'the suppression of the use of bhang would be totally unjustifiable. It is established that this use is very ancient, and that it has some religious connotation among a large body of Hindus; that it enters into their social customs; that it is almost without exception harmless in moderation, and perhaps in some cases, beneficial; that the abuse of it is not so harmful as the abuse of alcohol.'[7]

You could say that the Commission is (still) the only government report that has told the full story about cannabis. There hasn't been a single one in the years since that hasn't pretended that cannabis is more harmful than it is.

Prohibition propaganda

As mentioned in Chapter One, public fear of the effects of cannabis on mental health was used later on by Harry Anslinger, commissioner of the Federal Bureau of Narcotics and creator of the Drug Enforcement Administration in his campaign to ban cannabis after the end of alcohol prohibition. His tactic was to create hysteria around cannabis.

Basing his campaign on fear, he used racist propaganda – such as the film *Reefer Madness* – to convince the public that cannabis made people violent and insane. He rebranded cannabis, calling it by its Mexican name – marijuana – to more closely associate its use with illegal Mexican immigrants.

Working with the less scrupulous and anti-immigrant sections of the media, he created a racist narrative based on fantastic scare stories about the damaging impact of cannabis. Common themes included the idea that drug-crazed, dark-skinned immigrant men from Mexico and Africa would, under the influence of cannabis, rape white women and give the drug to white boys. 'Marijuana is the most violence-causing drug in the history of mankind,' he declared. His campaign led to medical cannabis being banned by the League of Nations (the forerunner to the United Nations), and eventually, in 1971, in the UK.

THE SCHIZOPHRENIA DEBATE

The modern debate on mental health, about schizophrenia and cannabis, started in the 1980s and has continued, in one form or another, until the present day.

The first study to link cannabis and schizophrenia came out in 1987. In 1957, a group of 17-year-old Swedish conscripts were asked if they'd smoked cannabis. Thirty years later, their records were followed up. It turned out those who'd smoked cannabis rather heavily were six times more likely to be diagnosed with schizophrenia in later life.[8]

At the time this study was well covered in the press, which reported it as proof that cannabis caused schizophrenia. This helped kick-start a stream of similar research in the area, and also created a lasting public perception that cannabis causes schizophrenia. Successive UK governments have used this to support their anti-cannabis stance.

Schizophrenia is a devastating disorder, diagnosed most often in men in early adulthood. It is difficult to manage and will often have a very damaging effect on the person's well-being and future life.

That's why the search for a cause that might lead to prevention or treatment has been a major theme of psychiatric research for the past hundred years. Doctors have suggested multiple causes over this time, none of which have stood up to research scrutiny. Cannabis has been one of the most enduring, in particular after the rise of its use in the second half of the twentieth century.

This isn't surprising. People who use a lot of cannabis can appear impaired in the same way as people with schizophrenia. We have known since the 1970s that THC can make people have psychotic experiences. Professor John Krystal's group at Yale have even used THC in human volunteers as a model of psychosis.[9] We know cannabis can give you psychotic experiences, including being paranoid or hearing voices.

It's a very appealing hypothesis for other reasons, too. Experts wanted to believe it, because it would give them a preventable cause for the disorder as well as clues on how to treat it. That's the reason the CB1 receptor antagonist rimonabant was developed – as a possible antipsychotic, although it turned out not to work.

The theory was, if you could stop people from ever smoking cannabis then maybe some cases of schizophrenia would never emerge. Parents wanted it to be true so they had something to blame for their child's illness – the drug, the dealer, even the child themselves for taking the drug. This last is an understandable reaction but one that's exceptionally unhelpful to the person with schizophrenia, because parental hostility (technically called negative expressed emotion) is known to worsen outcomes.

Behind the Swedish study

In fact, the Swedish study results did not claim to show causation, only correlation. The study showed that people who use cannabis are more likely to be diagnosed with schizophrenia. The study didn't look at other factors, such as that people with schizophrenia are also more likely to have used other drugs and to have smoked tobacco, as well as having problem behaviour at school, and to have stolen and lied. But none of these were pointed out as possible causes for schizophrenia, although cannabis was.[10]

What happens when we look at the issue from a population viewpoint? If cannabis does cause schizophrenia, as its use rose hugely in the latter part of the twentieth century you'd expect to see a corresponding rise in schizophrenia. For example, you can see this happening clearly for alcohol use and liver disease: a 50 per cent increase in the consumption of alcohol, as happened in the UK in the 1990s, more than doubled the levels of deaths from alcoholic liver disease.[11, 12]

But this is not what we saw happening for levels of schizophrenia. In 1970 in Britain, there were half a million people who'd ever used cannabis. By 2010 there were 12 million. That's a 24-fold increase in the number of people using cannabis (and likely a hundred-fold increase in the amount of cannabis used).[13]

But there was no increase in the number of people with schizophrenia. Or even in the number of people with psychosis in general. If anything, the number as a proportion of the population fell.[14]

The same holds for other Western countries where there's been a similar rise in the use of cannabis.

SCHIZOPHRENIA OR PSYCHOSIS?

There's a possibility that some of the Swedish cases in the original study may not have been schizophrenia at all, but cannabis-induced psychosis being mistaken for schizophrenia. These two conditions may look similar, but they are not the same.

Some people will have an acute psychotic reaction when they take cannabis. This has led to an increase in people presenting to A&E with psychotic symptoms. This kind of reaction usually resolves in a few hours.

But consistent use of cannabis can produce symptoms that look like schizophrenia, such as a loss of motivation, hallucinations, dissociation, hearing voices, paranoia and confused thinking.

The condition cannabis-induced psychosis is diagnosed when these symptoms are significantly impairing, last several weeks and can't be explained as complications from other drugs such as cocaine. This will happen to a very small percentage of people who smoke cannabis.

Within this group of people with lasting symptoms, some will have schizophrenia, but in others it will be the residual effects of the cannabis. If the person stops using cannabis, they tend to disappear within a few weeks or months.

Schizophrenia is a more complex disorder, with three components:

1) hallucinations and delusions
2) negative affect, loss of interest, motivation and volition
3) cognitive impairment that's something like dementia

People with cannabis-induced psychosis tend to have fewer negative symptoms, more positive symptoms, more anxiety, and more awareness that they are ill. Someone who's schizophrenic, on the other hand, will feel as if their brain is giving them insights that make sense of the world.[15]

When researchers scanned the brains of patients who were experiencing an acute psychotic episode, they showed cannabis had a different effect on the brains of the two groups.

It appeared that using cannabis increased brain metabolism in patients with psychosis but not in those with schizophrenia, showing that the two illnesses can be distinguished in terms of brain dysfunction.[16]

What we know now

Despite a further 20 years of research, the link between cannabis and schizophrenia is still unclear. But over the years, the blame has shifted. First, it shifted to high-THC skunk rather than cannabis causing schizophrenia. Then it shifted to skunk causing schizophrenia, but only in people who have a genetic disposition.

Current evidence suggests a much weaker link. It does show that people with schizophrenia are more likely to use cannabis. But this link appears to be correlative rather than causative.

In the last ten years, US studies have been able to look at the genetics of people who use cannabis, and we can see that a large part of the risk of schizophrenia is controlled by genes. There are non-genetic risk factors too, such as the mother having an infection during pregnancy, perinatal problems or a complicated birth; in fact any insult to the brain of a baby in utero or very early in life.

We now know that if you have a genetic predisposition to schizophrenia, smoking cannabis means your disorder will manifest itself earlier. The psychotic experiences of cannabis facilitate the emergence of your underlying disorder, in the same way as smoking does for a predisposition to heart disease.[17]

Studies have also found that the genes associated with schizophrenia are also associated with cannabis use. So if you have a genetic propensity to have schizophrenia, you have a genetic propensity to use cannabis. You are also more likely to use more of it, and earlier in life.[18]

It does appear that the rise of skunk is part of the problem. The balance of evidence from the UK now suggests that while the use of traditional herbal cannabis is not associated with psychosis, the use of high-THC skunk – and certainly spice – may worsen symptoms of schizophrenia in people who use it a lot.[19]

This is ironic. The prohibition of cannabis, fuelled by fear-mongering about mental illness, drove the rise of skunk. And, in turn, skunk has increased the amount of possible harm from cannabis, in terms of both dependence and psychosis.

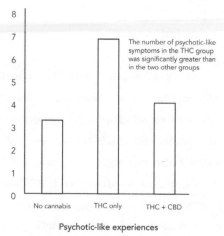

The number of psychotic-like symptoms in the THC group was significantly greater than in the two other groups

Psychotic-like experiences

Source: https://pubmed.ncbi.nlm.nih.gov/18378995/

Fig. 11 Psychotic-like experiences in users of different forms of cannabis

THE CANNABIS MENTAL HEALTH CONUNDRUM

Doctors report that patients with psychosis who use cannabis are less likely to recover than those who stop. If you are hearing voices, cannabis may make those voices louder so you're less able to cope with them, or you may even become more angry with them and also with other people.

For both schizophrenia and other types of psychosis, continued use does predict a worse outcome.[20]

This leads to an interesting question: why do people who are psychotic use cannabis, if it makes their symptoms worse?

Perhaps it doesn't feel worse to them. It may be a form of self-medication, relieving some of their other symptoms such as slowed thinking or anxiety. It may help reduce the adverse effects of their medication, such as muscle stiffness. Cannabis may help people sleep, or improve their appetite. It may distract them by increasing their enjoyment of activities, such as appreciation of music.

Many people with schizophrenia don't live in great conditions, and few are employed, so being stoned may simply be a way to get through life.

CBD as an antipsychotic

There's now some evidence that CBD is an antipsychotic. We know this from studies where it's been added to medication to treat schizophrenia and has improved patient outcomes.[21, 22]

Some clinicians are now prescribing people with schizophrenia Sativex to help them stop using skunk.

This makes sense, given the work at the Central Institute of Mental Health in Mannheim led by Professor Markus Leweke. It showed that CBD increased levels of anandamide in the spinal fluid and this was associated with a decrease in psychotic symptoms.[23]

This suggests the endocannabinoid system plays a role in schizophrenia, which may ultimately help us understand why people with the condition use cannabis.

WHAT TO TELL YOUR TEEN ABOUT CANNABIS

1 The later in life you start using, the less likely you are to have any resultant harms, including becoming dependent. Early use – in particular before the age of 15 – appears to be riskier.

2 If you are using cannabis, don't do it every day. If you start to use it habitually, that's when you may get into problems, in particular dependence. If you start to use bigger amounts, this may be a sign you're at risk of addiction too.

3 Avoid skunk, the name for high-THC plant. Smoking it raises your risk of psychosis and dependence. Don't smoke a lot and don't smoke every day. NB: You may find it hard to avoid skunk, as it makes up most of what's sold by dealers in the UK.

4 Never, ever use spice, the name for synthetic cannabinoids. It is not cannabis. It might also be called many names including X, Tai High Hawaiian Haze, Mary Joy, Exodus Damnation, Ecsess, Devil's Weed, Clockwork Orange, Blue Cheese, Black Mamba, Annihilation, Amsterdam Gold.

Some of these won't have been tested even on animals, let alone on humans. They can make people catatonic, violent, have seizures and heart attacks, become dizzy and collapse and/or vomit. (And, this is off-topic for this book, but it bears saying here: never use opiates such as heroin or fentanyl. And never inject any drug: injecting gives a profoundly powerful effect that can also be extremely addictive, as well as more likely to kill you and give you infections.)

5 If you are smoking cannabis, ideally do it without tobacco, for example in a bong or a water pipe. Never vape cannabis that's dissolved in solutions of unknown origin. In the US there have been a number of deaths from vaping cannabis oil dissolved in vitamin E oil (see the box on EVALI below).

6 Don't smoke and drive. Especially don't smoke and drink or take any other drugs and drive. Or do anything after smoking that you need a clear head for.

7 Monitor how cannabis affects you, to learn when you've had enough, in the same way as you would for alcohol.

What to do if your teen is smoking cannabis

In 2020, around 14 per cent of 16- to 17-year-olds said they'd smoked cannabis in the past year, so your teen's behaviour is relatively normal. Two-thirds say they use it only infrequently.[25]

Do not rush to the GP and especially not to the police. Approach the issue the same way as if they were staying out late and drinking, or breaking the law in another way: as a behavioural issue that you deal with, with sympathetic parenting. Just because cannabis is an illegal drug doesn't turn it into a legal problem.

CANNABIS

Cannabis doesn't kill people but evidence suggests it might be bad for a teen's developing brain. However, it seems this effect may only happen if 1) they smoke a lot and 2) they smoke strong cannabis, aka skunk (which does make up almost all of what's sold on the illicit market).

Digging down into the findings, we see that teens who use cannabis before the age of 15 have lower IQ scores and poorer educational performance. However, when adjustments are made for related factors such as behavioural problems and depression, and being more likely to use alcohol, cigarettes and other substances, using cannabis is not predictive of IQ or education. We only see effects on IQ when people keep smoking heavily into adulthood.[26]

Overall, the effects on the teen brain from cannabis don't seem nearly as dangerous as the harms from alcohol: fighting, violence, sexual and other types of risk-taking, accidents, being more vulnerable to attack, alcohol poisoning and overdose.

Parents often worry that cannabis is a 'gateway drug', which it is not (see Chapter Twelve). However, because it's illegal, it's often sold by dealers who also deal harder drugs, which means your child may be exposed to these. Keep telling them never to inject any drugs or to use any opiates such as heroin or fentanyl.

Telling your child you are concerned about their use of a drug – whether it's alcohol, smoking or tobacco – is not an easy conversation. You want to explain that it may not be good for their brain if they smoke a lot of cannabis, but at the same time you want to keep open the lines of communication.

WHAT ARE THE REAL HARMS OF CANNABIS?

- Know that for almost everyone, using cannabis will be a short phase of rebellion or having fun. Being very harsh or forceful is more likely to shut discussion down. Read this book and give it to them too so they can read what's true about cannabis. Discuss it together.
- Work out what you want and what your concerns are before you have the conversation. Do you want them to stop or would you be happy if they used it less? Do you want them to have counselling?
- What don't you like about them smoking? Is it because it's illegal, or because you don't like the lifestyle that comes with it, or you're scared of them becoming dependent?
- Are you scared they are damaging their future prospects or going to fail their exams? Have you had problems with alcohol or other drugs in the past?
- Try to find out and understand why they are using cannabis. Was it a one-off or is it a regular thing? Are they anxious or depressed? Are they scared of something? Does it help them in any way? Is there peer pressure so they feel coerced into taking it? If they are just smoking it for fun, they are much less likely to become addicted, the same as for alcohol.
- Find out if they might be addicted (see Chapter Twelve for some questions to ask). Do they want to stop? Can they stop? If they can't stop, your GP is a good first port of call. Make sure they know you are there to help if they do want to stop or cut down.
- If your child is displaying psychotic symptoms such as talking to themselves or being suspicious of other people, then persuade them to stop and see if it resolves. If it

doesn't stop or they won't stop, take them to talk to your GP. GPs do not go to the police.

– Tell yourself and them that it's not your child's fault if they smoke and become psychotic or dependent. What has happened is due to them being vulnerable to these effects. Blaming them for smoking won't help. In fact it can stop your child getting better. We know from research that the way families express emotion is a major factor in developing and perpetuating psychosis. This includes negative emotions such as blaming your child for their illness and being hypercritical, and also seemingly positive ones such as being overprotective and making excessive emotional requests.

– One thing you should never do is take your child to the police. The law is a blunt and very dangerous instrument in dealing with this situation. The resulting damage to a young person's life from a criminal record is much greater than from the cannabis itself. They may be kicked out of school or college or lose their job. They will not be able to travel to certain countries including the US. And they will have a criminal record that could severely limit their career opportunities.

– See Drugscience.org.uk, a good source of information on all these topics

SKUNK

Skunk differs from traditional cannabis in two ways. It is stronger as it contains more THC (the active 'stoning' ingredient of cannabis), around three times that of traditional cannabis (on average 15 per cent versus 5 per cent).

Also, when the plant is forced to grow more THC, it makes less of the other main ingredient, CBD. This is a non-psychoactive ingredient that may have medicinal properties (see Chapter Eleven) but which also seems to act to moderate some of the more extreme effects of THC.

As you've read, you're more likely to become dependent on skunk than traditional cannabis, and skunk is more likely to produce psychotic effects, too. It's thought CBD may help reduce THC's pro-psychosis actions and make it less addictive too.

If you want to keep using cannabis, it would be better for you to source a lower-THC, higher-CBD herbal cannabis, as this seems to have fewer risks. That said, it's very hard to buy on the illicit market.

If we had a regulated market, you would have more of a choice around the strength of the cannabis you bought. The government could discourage the use of skunk by pricing cannabis according to the THC concentration. Taxing according to strength has worked for controlling the consumption of strong alcohol for the past 300 years in the UK.

The Dutch approach has another way of minimising this risk; most cannabis is purchased in cannabis cafés, where the THC content is known and advertised and there are advisors on hand. The average UK user of cannabis has virtually no idea what strength of cannabis they are getting, so is much more likely than their Dutch counterpart to accidentally take too much. However, the remarkable safety of cannabis means that not even skunk is lethal.

> It may be that you just want to get as high as possible, as fast as possible. That is what skunk was designed to do. Compare this to alcohol. Everyone knows that drinking spirits by the pint is considerably more dangerous than drinking beer by the pint. The same is even more true for skunk as compared to herbal cannabis, due to the absence of CBD. If you don't want to give up, you could try taking CBD as well.

Accidents and suicide

Just like with alcohol, you shouldn't drive or operate machinery if you are intoxicated (more on driving and cannabis in Chapter Fifteen).

A Canadian study analysed three years of data on accidents at work. When the researchers compared people who reported using cannabis to those who hadn't, they found no association between past-year cannabis use and work-related injury.[27]

Poisoning

After legalisation, there is an increased risk that children may come across cannabis and take it. A report from Michigan, where medical cannabis became legal in 2008, showed 428 instances of under-18s having their cannabis use reported to the Michigan Poison Center between 2008 and 2019. Of these, 327 were from ingestion and 79 from inhalation. The latter tended to be teens (and perhaps are less likely to be accidents).[28]

This isn't unexpected. The more a substance is around at home, the more likely children are to find it and take it. And it's especially predictable when you consider that THC gummies often come in fruit flavours, are rainbow-coloured and shaped like teddy bears.

However, there don't seem to have been any reported child deaths from cannabis toxicity. We know very high doses of CBD are safe in children with epilepsy. And we know the toxic dose of cannabis is extremely high when compared to the amount people usually take (see Chapter Ten).

Compare that to laundry capsules, another product that toddlers are tempted to eat. Contact with the chemicals in capsules can cause eye damage and burns, as well as breathing problems, seizures, coma and even death. There were 10,572 cases of accidental ingestion reported in the US in 2020, according to the National Poison Data System of the American Association of Poison Control Centers.[29, 30]

The lesson is a practical one. Just as you would keep medication, alcohol and laundry capsules away from children, keep any cannabis products away from them too. And in particular, keep gummies away.

Cannabis and cardiovascular events

In Chapter Seven, I described how cannabis can – rarely – cause an irregular heartbeat, as well as increasing heart rate and blood pressure.

Research suggests you're at higher risk of a heart attack and angina in the 24 hours after you've used cannabis, especially if you have a history of heart problems. This effect is rare but it can be life-threatening.[31]

The advice is, if you have a heart issue, be very cautious about using cannabis.

Fertility

Over the years, there have been a lot of scare stories about a link between cannabis use and decreased fertility. Measuring fertility is difficult, but population studies show no evidence of delayed conception. A US study of men and women who were trying to conceive found that the time taken to conceive, even for daily users, was the same as for non-users.[32]

That said, there are potential mechanisms for how cannabis could impair men's fertility, specifically in reducing sperm count and concentration.[33]

And research has found a relationship between cannabis use and abnormal sperm morphology.[34]

However, the research isn't clear: a fertility clinic study in the US showed that men who'd ever smoked cannabis had higher sperm concentration and sperm count.[35]

For female fertility, we know that cannabis receptors are found in the oviduct and uterus, and that endocannabinoids play a key role in female fertility. And we know THC can alter the menstrual cycle in animals.[36]

But we don't have enough evidence about cannabis and fertility to make clear recommendations. The usual advice that's given to couples is not to smoke tobacco while trying to conceive. Women are advised not to drink alcohol, while men are advised to stick to 14 units of alcohol unless they are having problems conceiving, when they're advised to stop. It seems sensible to keep cannabis advice in line with this and advise people not to take it.

WHAT ARE THE REAL HARMS OF CANNABIS?

Pregnancy

One of the main arguments used by those who oppose any change to the legal status of cannabis is that it will lead to more pregnant women using it. Similarly to alcohol, women may use it before they know they're pregnant; or they may use it because they are dependent.

In California, women are routinely screened for cannabis use by the insurance company Kaiser Permanente. This shows that between 2009 and 2016, the number of women using while pregnant went up from 4 per cent to 7 per cent. Some dispensaries in the US even offer cannabis products for morning sickness (nausea), although this is not advised.[37]

It's been known since the 1980s that THC can cross the placenta and, after birth, is found in breast milk, too. However, there isn't nearly as much research on what cannabis does to the developing foetus as there is for alcohol.

There is a suggestion that a high intake of cannabis is associated with pre-term birth.[38]

And it's thought cannabis use during pregnancy may increase the risk of the child having ADHD and depression in later life, although this is a theory rather than proven and the mechanisms aren't known.[39]

As with tobacco, the babies of women who use cannabis may have a lower birthweight, although it's not known if this is due to the cannabis or to smoking it with tobacco.[40]

The endocannabinoid system is clearly involved in many aspects of pregnancy, as it is with fertility. We don't yet know how cannabis might affect this. Until we know it's safe – if indeed it is – it's best to avoid it.

Cancer

See Chapter Eleven for a section on how cannabis may work against cancer. There is some evidence that high usage may lead to a small increase in testicular cancer.[41]

Studies on smoking cannabis and other types of cancer, such as lung cancer and head and neck cancer, don't appear conclusive, so more research is needed.[42]

The respiratory system

Smoking cannabis can lead to long-term damage to the respiratory tract and lungs, including raising your risk of chronic bronchitis and possibly lung cancer, although this is contested. However, it seems to be less damaging to the lungs than tobacco. This may be due to people smoking cannabis less often – studies suggest lung damage may only happen with very high use. Or it may be because cannabis burns at a lower temperature than tobacco. The anti-inflammatory effects of cannabis might also play a part here too.[43]

14

PROHIBITION AND THE RISE OF SPICE

HISTORY TEACHES US that trying to stop the use of one drug is likely to lead to the use of more potent, and more dangerous, alternatives. The international ban on opium smoking in 1912 led to the rise of heroin injecting, for example. And alcohol prohibition in the 1930s led to people dying from drinking methanol and home-made hooch. Crack use came about as a way to get more bang for your buck from cocaine.

This effect has also been one of the perverse consequences of the government's 50-year insistence on criminalising cannabis and persecuting cannabis users. Prohibition has encouraged the illicit market to make and sell stronger and so more dangerous versions of cannabis, namely skunk and spice.

I have described how skunk became the dominant form of cannabis on sale in the UK, and the negative effects of this on mental health and dependence (Chapter Thirteen). In 2009, when I was in the ACMD, the Labour government cited the increase in skunk use as a reason to move cannabis from class C to class B. What's ironic is, the government's prohibition on cannabis played a large part in the development of skunk.

The same is true of synthetic cannabinoids. Officially these are synthetic cannabinoid receptor agonists (SCRAs), but most people call them spice. They are man-made chemicals that bind to the same receptors in the brain as THC, but in a much stronger way. And it's this that accounts for their much stronger side effects.

I want to stress that spice is not cannabis. We should not blame cannabis for any of the terrible effects of spice.

Because spice is an even worse situation than skunk. The government's actions are actually now killing people.

You may remember when one kind of spice, called K2, hit the headlines as the cause of 'zombie' outbreaks in New York and recently in Manchester.[1]

There were shocking news pictures of homeless people, seemingly frozen in place, catatonic but standing.

Now, as well as on the streets, the other place that spice is rife is in prisons. This is due partly to it being cheap and easy to smuggle, but mainly – as I explain below – because it doesn't show up in urine tests.

WHAT IS SPICE?

On the street, spice is sometimes also called Black Mamba, but it's been sold under lots of brand names, including K2, Mary Joy, Devil's Weed, Ecsess, Amsterdam Gold, Annihilation.[2]

Spice comes in liquid form; it is soaked into a dried herb or paper, which is smoked.

Not all synthetic cannabinoids cause zombie-like loss of consciousness, but the side effects can include psychotic states, violence, fits (seizures), heart attack and respiratory failure.

Even the milder side effects are unpleasant: anxiety and paranoia, vomiting, confusion, amnesia.

The full range of unwanted effects is described in the paper by Dr Dima Abdulrahim and Dr Owen Bowden-Jones on behalf of the NEPTUNE group of which I was an author. There's a short version on page 214.[3]

Synthetic cannabinoids were first made in the 1970s as potential medicines, but initial human testing found them too unpleasant and potent to be marketed.

Over time, though, they did begin to go on sale. They were probably appealing because they were not specifically banned by the Misuse of Drugs Act 1971, so could be sold – along with many other compounds – in head shops as 'legal highs'. Another reason is that they don't show up on drug tests for cannabis.

WHY SPICE IS SO DANGEROUS

Firstly, there's no quality control: you cannot know the amount of synthetic cannabinoid in any dose, or be certain which chemical you're taking, even if the packaging looks the same as you've used previously. So you can't predict the effects – or side effects.

Secondly, these substances have little, if any, safety data.

Thirdly, many are much more potent than traditional cannabis, up to a hundred times in the test tube. They give a bigger effect and last for longer than cannabis, are more addictive and are more likely to cause withdrawal.

The horrible side effects have seen spice become less popular in the general population. But it's still used by homeless people and in prison because it's relatively cheap but very high-potency. As it can be soaked into pieces of paper, it's easy to carry and to smuggle into prison. It may also be that prisoners and the homeless are less bothered by the drug's profoundly dissociating and depersonalising experience, because it provides a form of escape from the difficulty of their lives.

The prison spice epidemic

It's not surprising that prisoners use drugs. Conditions in prison can be grim; prisoners can be locked in their rooms 23 hours a day, denied exercise, work and rehabilitation. Many have underlying mental illnesses. In the past, a lot of prisoners used weed to deal with prison conditions or mental ill health, to calm down and help them sleep.

The prison drug testing scheme was set up in 1995 as a pilot, as a research tool to find out about use in prison. It was tried in just a few prisons. Then, for reasons that are unclear, it was rolled out in prisons across the country, rebadged as a policy to reduce drug use.

Prisoners who tested positive were denied probation, which could increase their sentence length by several years.

Ten years later, during a review of the testing policy, it became obvious the policy was doing more harm than good.[4]

Prisons used urine tests, which, rather than testing for THC itself, detect a long-lasting metabolite of THC, which can stay in the body for up to ten weeks after use. At first, cannabis was the most detected drug. But when prisoners began to test positive for cannabis and be punished, its use fell as they replaced it with drugs that don't last as long in the body and so are less detectable, including heroin.

Testing was implemented against the wishes of some prison governors, who predicted it would make things worse. But the government chose to ignore the early warning signs that testing was causing more harm than it was preventing.

This focus on punishment of prisoners and drug users by government is typical. They want to be seen as hard on both drugs and crime. They can't be seen to be 'soft' in any way – such as allowing prisoners to smoke cannabis.

Then, in around 2013, spice arrived in prisons and quickly became one of the drugs of choice.

One prisoner said:

'. . . *They are using it because it's cheap, it's strong and because those who are out on licence will not go back to jail if they are caught taking them because they're legal.*'[5]

Spice is profitable for a dealer too. It's thought that up to 5 per cent of prison officers smuggle drugs.[6]

A solution costing a few pounds can be soaked into a single A4 sheet of paper, which, when dried, can be cut up into about 100 units, each of which will give a 'hit', for £5 each.[7]

Peter Clarke, the then HM Chief Inspector of Prisons, wrote in his 2015/6 annual report that spice 'caused major problems in most adult establishments that we inspected, including medical emergencies, indiscipline, bullying and debt.'[8] Prisoners report becoming addicted to spice while in prison, which reduces their chances of rehabilitation further.

British prisons now need teams of paramedics to deal with the violence, fitting and cardiac situations caused by spice. These health emergencies not only have a terrible human cost but also consume hours of prison officers' and health professionals' time, and so waste a huge amount of public money. Between 2013 and 2016, there were 64 deaths in British prisons linked to spice, including two murders.[9, 10]

The effects of this epidemic of spice have repercussions beyond the prisoners, too. Repeated demand for medical interventions has in some towns depleted community ambulance teams. One nurse reported having to treat over 50 prisoners for the effects of spice in one week, putting herself at risk in the process. Prison officers and nurses have had to go off sick after inhaling smoke from spice or even from touching the envelopes that the drug-soaked paper is sent in.[11]

There are tests for spice, but they aren't widely available and can't detect all the variants. And drug testing equipment is expensive, as are ambulances and staff.

PROHIBITION AND THE RISE OF SPICE

The government response to spice

I am not sure if the continuation of testing was a decision – it might just be that no one thought to stop it. But experts have, for decades, been saying that testing is damaging.

Drug Science has spoken with prison governors, legal experts and the police to try to explain that the spice epidemic was an inevitable result of the clampdown on cannabis users and in particular, cannabis testing in prisons. The governors understood but the police didn't seem to care. The government's approach was, as usual, to ban and keep on banning.

The government took the advice of the ACMD, and attempted to make all synthetic cannabinoids illegal by adding them, as controlled drugs, to the Misuse of Drugs Act 1971.

The first stage of this process was to ban compounds that were on the street at the time, most of which were the original cannabis substitutes – the first-generation compounds. That meant they had at least been tested on humans and we had some knowledge of their dose and effects.[12]

What did the dealers do? They went back to the chemistry journals and found compounds that had been made as follow-ups to these molecules but which had never been given to humans – the second-generation compounds. These turned out to be even more toxic than the first generation.

The ACMD then had to play catch-up, and they banned these in 2013.[13]

So the underground chemists went back to first principles and started making a whole range of completely new third-generation compounds that had never been tested on anyone or anything before. Predictably, these proved to be yet more toxic. As there are so many different chemical structures, the ACMD is still struggling to come up with legal definitions to ban them. Their first attempt was a disaster. The wording of the planned legislation would also have banned a whole range of medicines that have similar chemical structures but no impact on the cannabis receptor.[14, 15]

One major UK drug company estimated that in its library of compounds it had over 80,000 molecules that would have become immediately illegal if that legislation had been put through. Thankfully, organisations including Drug Science as well as the British Pharmacological Society pointed this out to the ACMD, who put a hold on the legislation. It is still under consideration.

In 2015, the government also announced a 'crackdown' on spice in prison, with new punishments for smuggling. Anyone caught could be denied contact with visitors, have their privileges removed and have time added to their sentence.[16]

Where are we now? Currently, there are multiple synthetic cannabinoids on the streets and in prisons that may not be controlled under the Misuse of Drugs Act. It is possible that they may be controlled under the recent Psychoactive Substances Act 2016. However, this act doesn't specify any compounds.

The upshot of the act is that every substance found by the police and prosecuted under the Psychoactive Substances Act requires a jury in a crown court to decide if the molecule is psychoactive or not. There have been very few cases where this has happened, which may explain why spice is still rampant in our prisons and on our streets.

Perhaps because the system is failing so badly, the Department of Justice now doesn't present data on spice.[17, 18]

You can find out, but only by looking carefully. Excluding psychoactive substances, 10.4 per cent of random mandatory drug tests were positive in the 12 months to March 2019. Including the psychoactive substances, the rate was 17.7 per cent. This shows that around 7 per cent of prisoners take spice.

What's the solution to spice?

First, the government must recognise that this problem is not going to go away with more bans or more severe sentencing. The main reason for the rise of spice in the UK is prohibition of herbal cannabis.

The then Police and Crime Commissioner and interim mayor of Manchester, Tony Lloyd, described how the situation was, in fact, made worse by the Psychoactive Substances Act 2016. This took synthetic cannabinoids out of 'head shops' and into the underground marketplace. 'The government didn't really think through the consequences of banning spice and criminalising the supply of spice,' he said. 'It should have been thought through as to what the impact would be if it led to less control over supply.'[19]

Perhaps we could reduce harms by allowing the sale of the safer synthetic cannabinoids, or even cannabis itself, in head shops.

But most urgently, in the short term, we need to stop testing for cannabis in prisons. This should hopefully encourage users back to herbal cannabis. Cannabis doesn't kill, but spice does.

*   *   *

We also need to research antidotes to spice. Naloxone, the antidote for heroin overdose, is now proven to save lives. But the government has failed to act to develop one for spice.

One possible class of drugs to use would be the cannabis antagonists, which block the spice brain receptors (see Chapter Six) and so are likely to stop or reduce its effects.

One in particular, rimonabant, which we have discussed, already has extensive safety data in humans. It was previously licensed in Europe as a weight control treatment but was taken off the market because of its negative effect on mood. But that shouldn't prevent it being tested as an antidote to a bad spice reaction.

The other approach is to develop the herbal antidote THCV, a cannabinoid from the plant. It's recently been shown to partially block psychotic symptoms promoted by THC.[20]

We don't know if it will work against spice, but it should be tested.

Sadly, the Advisory Council for the Misuse of Drugs refused to make it available for this purpose on the rather thin grounds that a note in a 1974 paper said that when given intravenously in high doses it was a little like THC. They ignored studies that showed it wasn't psychoactive when taken orally.[21]

I was highly critical of this decision, which I saw as just one more example of scientists erring on the side of prohibition 'just in case' without seeing the bigger picture.[22, 23]

I wrote to the health secretary Jeremy Hunt to stress that developing spice antidotes is a matter of urgency. I thought that even if the government weren't interested in this for humane and health reasons, surely they might consider it for the money it would save.

Their replies revealed a complete lack of appreciation of the size of the spice problem and lack of interest in an antidote. It seems the government has no interest in harm reduction or in the health of drug users. Again, the fixation on keeping things illegal trumped common sense, scientific evidence and medical need.

Recently I did an interview on spice in prisons on Radio 5 with presenter Stephen Nolan, sparked by the then North Wales Police and Crime Commissioner Arfon Jones, suggesting a trial of giving prisoners cannabis in order to reduce other drug use and violence.[24]

The conversation included Conservative ex-food minister Edwina Currie. As we spoke, it became clear to me that no evidence could convince either Nolan or Currie that drug testing in prisons was making the drug situation worse. When I asked if any data might make them consider reviewing the testing policy, they changed the subject.

This was not the first time I had experienced politicians and opinion-formers denying evidence, but it was the first time I had come across a complete refusal to discuss whether current policies were working. Without this kind of evaluation, we are on a treadmill of failed drug policies.

It's the same story all over again. Those who state they are 'against drugs' treat drug use as a moral issue rather than a society and health issue. And so they confuse harm reduction with condoning drug use.

CANNABIS

SIDE EFFECTS OF SPICE[25]

ADVERSE EFFECTS
Acute: Convulsions, hypertonia, myoclonus, wide-ranging cardiovascular effects including myocardial infarction and ischaemic strokes, acute kidney injury, hyperglycaemia, hypoglycaemia, vomiting, transient loss of vision and speech, reduced levels of consciousness, anxiety, aggression, extreme bizarre behaviour, amnesia, confusion, panic attacks, inappropriate affect, auditory and visual hallucinations, paranoia, delusions, psychosis.
Chronic: Psychosis, cognitive impairment, catatonic states, dependence, persistent vomiting, withdrawal symptoms on reduction or cessation of use.

WITHDRAWAL SYMPTOMS
Gastrointestinal cramps, nausea, tremor, hypertension, tachycardia, coughing, headache, craving, anxiety, restlessness, irritability, depression and suicidal ideation.

15

TESTING, DRIVING AND ACCIDENTS

WE ALL AGREE that intoxicated people – whether intoxicated by cannabis, alcohol or any prescribed or illegal drug – shouldn't be on the road. They should not be in charge of large machines that are capable of killing themselves and others.

However, those who are against changing the law on cannabis have used the issue of drug-driving as a reason to continue prohibition. They argue that allowing people to use cannabis would lead to a rise in the number of people drug-driving, and so to more accidents and more deaths.

For example, when Gordon Brown's government made the decision to move cannabis back from Class C to B, he cited driving as one of the three key reasons informing that decision, alongside skunk and schizophrenia.

WHAT'S THE UK LAW NOW?

In March 2015, cannabis was one of the 16 drugs included in a new drug-driving offence that was added into the Road Traffic Act 1988.

The stated aim of the legislation was 'to reduce expense, effort and time wasted from prosecutions that fail because of difficulties proving a particular drug impaired a driver.'

It made it illegal to drive with a controlled drug in the body, above that drug's accepted safe blood limit. Previously, you had to be impaired to be prosecuted.

This introduced a new kind of crime in the UK, one that's not in the Misuse of Drugs Act 1971: the offence of having a controlled drug in your body. It effectively creates this by the back door; if you are on or above a specific (and relatively low) limit while driving then you are automatically guilty.

In the new regulations, the government included cannabis as one of the eight drugs 'most associated with illegal use'. They described the newly created blood threshold for these drugs as 'zero tolerance'.[1]

This stated 'zero tolerance' approach to cannabis and other controlled drugs doesn't seem to have deterred people from driving under the influence. In England, police data showed that between 2017 and 2019 the number of convictions for drug-driving quadrupled.[2]

So what is going on? The government's blinkered focus on prohibition over safety on the roads is stopping them being able to communicate the real risks of drug-driving to people. Absolutist laws like this one stop people – who already know the drug is illegal – from listening and put a barrier up to creating a responsible culture around cannabis and driving.

The zero tolerance approach taken to drugs in these regulations is very different from our approach to alcohol and driving in Britain. We allow people to take much greater risks on the road after drinking; we have one of the highest allowed blood alcohol limits in the world. And alcohol, as you'll see below, has a much worse effect on driving ability.

Now that cannabis can be prescribed as a medicine in the UK, it's even more important to understand and communicate when it's safe to drive and when it's not. While we want to make sure that there aren't intoxicated people on the roads, we don't want to scare medical users into not using their cars when they are fit to drive. Or worse, to prosecute them for driving with very low and so inactive levels of cannabis in their blood.

DRUG-DRIVING PROCEDURE

The invention of the roadside cannabis saliva test allows the police to test drivers easily and quickly, and paved the way for the 2015 legislation.[3]

The police can now stop a vehicle if they have a reasonable suspicion the driver has taken drugs or is impaired, as well as test you after an accident.

Then they can ask you to do a breathalyser test, a saliva test and/or an impairment test. Since the 2015 change in the law, the police don't need to prove that your driving is affected by drugs in order to arrest you. They need only prove the presence of drugs in your system.

The 2015 legislation also gave them the power to require a sample of saliva. Refusal to provide this is an arrestable offence.

The test is a plastic stick that looks something like a pregnancy test. Once you've done the sample by swabbing it around your mouth, the officer will crush the vial of detector chemical in the holder to release it. Also like a pregnancy test, there's a control line to show the test is working, and then two more lines that show up if you've taken cannabis or cocaine (currently the only two drugs tested for at the roadside).

A positive saliva test isn't evidence of a criminal offence. But it means you can be taken to the police station for a blood test, which can be used in evidence. If your blood levels of THC are at or above 2 microgrammes per litre (mcg/L), you can be convicted of being unfit to drive through drugs. Even if your driving isn't impaired, you can be prosecuted for being over the legal limit.

You can get a criminal record, and a driving ban for at least a year. You could also be given an unlimited fine and up to six months in prison. The conviction will stay on your licence for 11 years.[4, 5, 6]

According to the manufacturer of the saliva test DrugWipe, D.tec International, all England, Scotland and Wales police forces now use their test. It is, they say, 'extremely sensitive', 'in just five minutes able to detect invisible traces of illegal drugs from any surface, liquid or substance, for example, cannabis at a level of 5ng/ml [5mcg/L] is a concentration equivalent to a teaspoon of sugar dissolved in an Olympic-size swimming pool.'[7]

TESTING, DRIVING AND ACCIDENTS

Saliva tests pick up any THC residue in the mouth of anyone smoking or taking an oil orally, rather than testing for the amount you've actually taken. They can detect THC for 12 hours and possibly for up to 72 hours after smoking, way after it would have stopped having any effect on driving. The detection window may even be longer in people who smoke cannabis regularly.

Saliva tests are currently causing huge controversy in Australia, where police are doing random testing at the side of the road, stopping and testing drivers even if the police have no reason to suspect that their driving is impaired. They are using a second sample of saliva taken at the roadside in evidence to prosecute.

In order for saliva testing to be an effective deterrent to drug-driving, it needs to accurately reflect what people have taken. A recent Australian study that assessed the accuracy of saliva tests showed a large number of false positives and negatives. It also showed that the saliva test may show up as positive if someone has taken very low-THC products, at levels similar to those found in CBD products on sale in the UK at high street stores. There's also a possibility of getting a false positive from passive intake; that is if you've been in the room with someone who's been smoking cannabis.[8]

Canadian snowboarder Ross Rebagliati won a gold medal in slalom at the 1998 Olympic Winter Games in Nagano, Japan. But three days later, he had his medal removed by the International Olympic Committee (IOC) because his drug test for cannabis had come back positive, albeit at a very low level. He was accused of importing a controlled substance and locked up. He said that he hadn't smoked for ten months, and the positive result must have been caused by second-hand smoke at a party his friends threw him before he left for Nagano. He was released. Eventually he had his medal returned, not because of his passive smoking defence but because cannabis wasn't a substance banned by the IOC. His story began a debate about cannabis in Canada that eventually led to decriminalisation in 2018. Ross is now a cannabis entrepreneur.[9]

HOW DANGEROUS IS DRIVING UNDER THE INFLUENCE?

Now, let's look at the real risks of driving after taking cannabis. We have known for 50 years that THC intoxication can affect motor skills. We also know there are CB1 receptors in brain regions that are in charge of cognitive and motor control. However, so far it has been hard to establish the real effects on the roads.

Some of the first research into this took place on airline pilots in the 1970s. It was set up after the authors had been 'informed of a number of instances in which pilots have flown aircraft while "high" on marihuana [sic].' They recruited six pilots, gave them either THC or a placebo to smoke, then put them in a flight simulator and asked them to do manoeuvres. As predicted, they found some short-lived impairment.[10]

The authors did a similar study two years later, where they gave ten pilots who had experience of cannabis socially either a placebo or cannabis to smoke. All showed a decrease in flying performance 30 minutes after the cannabis, and this continued for two hours.[11]

Since then, research done in labs and driving simulators has shown us that THC dulls people's concentration and attention, their reactions are slowed, their manual dexterity and coordination decline, and they have distorted distance perception.

But how does this translate to the road? In the real world, there's not a lot of evidence that an increase in people using cannabis leads to an increase in road traffic accidents.

Population studies on drivers testing positive for cannabis show that the risk of an accident is increased if you have any cannabis in your body. As a general rule, the higher your THC blood level, the greater the impairment. However, compared to alcohol, your risk is not massively increased.

In 2012, the Department of Transport asked Kim Wolff, Professor of Addiction Science at King's College London, and a team of academic experts, to look at the issue of drug driving.

Her 2013 report states:

'Impairment by drugs was recorded as a contributory factor in about 3% of fatal road accidents in Great Britain in 2011, with 54 deaths resulting from these incidents. This compares to 9% or 156 fatal road incidents, with 166 deaths, which have impairment by drink reported as a contributory factor.'

The report concluded that, on average, if you are driving with cannabis in your blood, you are almost twice as likely to have an accident. The risk if you have alcohol in your blood is eight times. And if you've taken both, the risks increase.[12]

In a more recent analysis review of crash risk literature, the average risk of being involved in a crash if you've taken cannabis was scaled down to 1.3 times. Even this may be an overestimate because it doesn't take into account the possible biases, namely that more young men are more likely to smoke cannabis and to drive at night and so to have accidents.[13]

It's possible that using cannabis at a low level may not increase the risk of an accident at all. Especially given that cannabis reduces impulsivity and if you are stoned, you're less likely to drive (see below for more on this).

WHAT HAPPENS WHEN CANNABIS IS LEGALISED?

Recently, we've been able to see from other countries and territories what happens on the road when we take away the illegal status of cannabis. These places are acting like a controlled experiment; you can compare before and after, and between similar places too.

Some initial data from Colorado showed that after legalisation, there was a small increase in people being stoned and having driving accidents.[14]

However, other reports have suggested there is an overall decrease in fatalities on the road.

For example, when Colorado was compared to Oklahoma – where cannabis is not legal – there were similar changes in cannabis-related, alcohol-related and overall fatality rates on the roads.[15]

It may be that some people switch from using alcohol to using cannabis, and this leads to a net reduction in road traffic accidents.

This may be down to fewer people driving when impaired. Being stoned doesn't tend to make people want to go anywhere, so people who've taken cannabis have less incentive to drive.

Or it may be down to differences between driving under the influence of cannabis and of alcohol.

Drunk drivers lack judgement as to how impaired they are. Alcohol disinhibits and dampens your sense of fear, so drunk drivers take impulsive risks and can stop caring about breaking the law. People under the influence of alcohol often think they are better drivers than they are.

On the other hand, stoned drivers know they're impaired. If they do drive, they may drive more slowly.[16]

They are less impulsive, so may be less likely to drive dangerously in ways such as getting into a race with another car or jumping traffic lights.

WHAT SHOULD THE LEGAL THC BLOOD LEVEL BE?

As a society, we agree there should be a maximum limit for blood alcohol in order to reduce accidents caused by impairment. The blood level limit is a measure of how impaired the government thinks it's acceptable to be on the roads.

In England, Northern Ireland and Wales, our level is one of the highest in the world, at 80mg of alcohol per 100ml of blood (80mg%).

Other countries have decided on a lower limit. Most of Europe is at 50mg%, and Norway and Sweden are at 20mg%, effectively zero tolerance for any drinking at all.

When the Department of Transport first looked into

establishing blood limits for THC, the focus was also on finding a limit for how impaired a driver could be.

In her 2013 report, 'Driving Under The Influence', Professor Wolff describes how her group was tasked to establish whether it was possible to identify for various drugs – for an average adult – 'blood concentrations equivalent to a blood alcohol content (BAC) of 80mg%'. And also: 'To establish whether in some specific circumstances different concentrations of these drugs (broadly equivalent to a blood alcohol content (BAC) of 50mg% and 20mg% may be deemed necessary for road safety.'

Wolff and colleagues looked at the best evidence, and for cannabis recommended a level of 5 nanograms per ml (5ng/ml, also written as 5mcg/L). The report said: 'It is noted that significant increased accident risk was apparent when the concentration of THC in the blood was 5ng/ml whether or not ingestion had occurred recently and regardless of the origin of the drug (medicinal or illicit).'

The report also recommended a dual limit, for alcohol and cannabis, recognising that the effects are additive. The recommended combined limit was 3mcg/L for THC and 20mg% for alcohol.

Instead, the government decided to use a blanket limit of 2mcg/L for THC.

It's good to err on the side of caution. But the experts recommended a higher limit, then government deliberately set the blood level for THC way below the level that is known to produce impairment. This is comparable to setting a breath test for alcohol at zero. The reason they did this was to deter drug use by punishing people for taking drugs, rather than punishing people for impaired driving. And their grounds for doing this is that cannabis driving is dangerous, when there's very little evidence.

The 2mgc/L limit isn't logical, given that the legally allowed blood alcohol limit allows a far greater risk of accident.

It's also illogical when you consider how the 2015 regulations decided on blood limits for other medical drugs, such as some opiates and benzodiazepines. The limits for these other drugs were based on 'road safety risk', in the same way as alcohol, not 'zero tolerance' as for cannabis. Since medical cannabis was made legal in 2018, there's now a strong argument that its limits should be based on 'road safety risk' too.[17]

How does the 2015 British THC limit compare to other European countries? It's higher than 1mcg/L in Belgium, Denmark, Ireland and Luxembourg. In Norway, the level for prosecution is 1.3mcg/L, although there are also higher penalties at levels of 3mcg/L, then 9mcg/L. It's the same as the Czech Republic. And in the Netherlands, where cannabis isn't illegal, it's 3mcg/L (for THC only). It's notable that this last is the only country that has a higher limit.

Deciding the right cut-off blood level is complicated. With alcohol, there's an established correlation between a person's blood alcohol level and their level of impairment.

This isn't true of THC blood levels, although we do know the bigger the dose of THC, the bigger the impairment. Research suggests a THC level of 3.8mcg/L is similar in impairment to a blood alcohol of 50mg%.[18]

We know CBD doesn't affect driving.[19] And we know roughly how long THC intoxication lasts. These graphs in figure 12 show the time course of blood concentration of a 10mg dose, which is high for casual use. The average spliff is 7mg. Most people will share that, so may take in 3mg.[20]

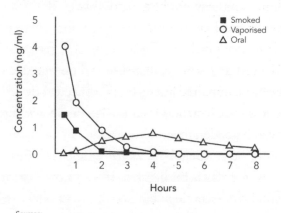

Sources:
Oral: https://www.ncbi.nlm.nih.gov/pmc/articles/PMC5890870/
Smoked:https://jamanetwork.com/journals/jamanetworkopen/fullarticle/2716990

Fig. 12 Time course of blood concentrations following different routes of administration of 10mg THC

As you can see, the greatest effects and highest blood levels are 20 to 40 minutes after smoking, and they then taper off over four hours.[21]

However, vaping gives you 50 per cent higher blood levels at ten minutes than smoking the same amount in a joint.

And taking it orally gives you a much slower and lower 'high', but the highest point happens at four hours, then it tapers off slowly.[22]

The question anyone who uses cannabis will ask is, how much will take you over this low limit? If you smoke a whole spliff of 7mg THC, even at peak blood levels you may not exceed the 5mcg/litre threshold, but you will exceed the 2mcg/litre threshold.

As a fat-soluble substance, cannabis hangs around in the

body and in the brain longer than it does in the blood. Blood levels fall off faster than the brain effects, so it's best not to drive for four hours. Even though you might not exceed the blood limit, you may not be safe to drive.[23]

However, there are big variations in impairment that aren't to do with how much you've taken or how recently. If you're inexperienced at taking cannabis or you don't take it much, you'll be more impaired at any blood concentration. Tolerance leads to less impairment. Other factors make a difference too: your age and gender, metabolism, body composition and size.

If you've been a regular and heavy user, THC can stay in your body for days. Even if you then become abstinent, it might still be detectable in blood.

On the other hand, the saliva test may stay positive even when you've recovered from impairment. Levels are often still detectable in saliva tests after 24 hours and even up to 72 hours (see box on page 229).

This is a complicated situation not only for the driver, but also for the police. Around a third of people who have a positive saliva test at the roadside turn out to be under the still very low 2mcg/L limit. Many police forces have reported finding this disappointing, which has led to some questioning the value of the current procedures.[24]

Perhaps the best way forward would be to include a measure of intoxication in the assessment, as some US states do.

What does this mean if you take cannabis for medical reasons?
The advice is the same for all medications, legal and illegal: if you feel intoxicated, do not drive and always be clear on whether or not you are over the legal limits too. Do not drive if you are.

One of the reasons for the 2015 legislation was that the Department of Transport was concerned there would be an explosion in the use of medical cannabis.

However, if medical cannabis were made more affordable and available, it's more likely that the hundreds of thousands of people using street cannabis for medical reasons would switch to lower-THC and higher-CBD legal versions. And as most people taking cannabis for medication try to avoid getting high or impaired, the result would be less intoxication on the roads rather than more. As above, being practical about the fact that people take cannabis and being open with them about the risks will allow us to educate them on when it's safe to drive, too.

Current evidence on psychological impairment suggests it may be safe to drive four hours after taking cannabis medicines, the same time period as recreational cannabis. But this does not mean you will necessarily have a blood cannabis concentration below the legal limit, so you could still fail a roadside test. A review of studies that looked at impairment found no difference between the placebo and medication group after that time, however big the dose of THC.[25]

That said, you are also more likely to take medication in oil form, which lengthens the time until the peak of the effect as well as the duration, and may lengthen the impairment to six to eight hours.[26]

This is not clear from current studies, as the four-hour limit was based on inhaled THC.

If your medication does contain THC and you take it frequently (defined as more than four times a week), you will likely be less impaired at driving than someone who has just taken the same dose, but isn't used to it.[27]

However, there are certain conditions where you may already have a higher risk of having an accident, for example multiple sclerosis, chronic pain or epilepsy, as well as insomnia, anxiety and depression, due to symptoms such as fatigue, weakness, slowed thinking or dizziness. The other side of this is, if cannabis gives you good symptom management, your driving may even improve.[28]

The legislation currently doesn't make an exception for anyone who's taking cannabis as medication.

If you're stopped and asked to take a saliva test, it's likely to show up as positive.

When Professor Wolff recommended a 5mcg/L limit, she said this would likely not take anyone with a prescription over the limit: 'The threshold is higher than average concentrations observed in those prescribed the drug.'

But this is not certain for a limit of 2mcg/L.[29]

If you've had an accident or are obviously impaired, if it can be proved that you were impaired then you can be prosecuted whether you're over the limit or not (Section 4 RTA). If you aren't impaired but you are over the limit you can still be prosecuted, but you will have a statutory defence of using cannabis for medical reasons (Section 5A RTA). This says that you're not guilty if the medicine was prescribed, supplied or sold to treat a medical problem and you took the medicine according to the instructions given by a prescriber, pharmacist or on the packet. But if you are impaired this defence will not hold, so it is important to monitor your mental state before driving.

I advise that you raise this fact at the earliest stage possible, i.e. with the policeman at the roadside. You could carry a copy of your prescription and/or a letter from your prescriber or your CANCARD card.[30]

If you do not have a prescription and so are using street cannabis for a medical condition, you could apply for a Cancard (see page 123). This might help you, but it's at the discretion of the police or, if they take the case further, of the Crown Prosecution Service.

The Cancard website says the card will 'prove that you are legally entitled to a cannabis prescription and are only in contravention of the Drugs Act because you are unable to afford one.'[31]

FOR HOW LONG IS CANNABIS DETECTABLE?

With the rise in drug use has come a rise in testing at work. You could argue that this is a grey area: what about if someone has gone to Amsterdam or somewhere else where you can buy cannabis legally for a weekend, and smoked some joints but not broken any laws; and has then come back, got tested and lost their job? And what about if someone has taken medical cannabis?

In the UK, employers can say it's illegal to have cannabis in your blood unless you have a prescription. And employment law may allow them to dismiss anyone if they've got something in their body they shouldn't have. You might be able to appeal at a tribunal and may even win, but that would be expensive and time-consuming.

As THC is fat-soluble, it is stored in your fat cells and so can show up in your system for some time afterwards, especially for regular users. The more body fat you have, the more THC you will store. And the more you take and the more regularly you take it, the more likely any test will show up as positive.

If you do fail a drug test, you could find out if it measured THC or a long-acting THC metabolite. If it's the metabolite, you could argue that it didn't have any effect on your function.

You could also point out that it's not illegal to have taken an illegal drug in Britain. It's only illegal to possess one or, if driving, be impaired by it.

* **Urine test**
  For non-regular users, THC is detectable in urine for three days. For more regular users, it may still be detectable up to 30 days later.
* **Saliva test**
  For 24 hours, but up to 72 hours for more sensitive tests.
* **Blood test**
  This is what is used in evidence by the police. THC doesn't last as long – three to four hours. But if you are a regular, heavy user, you may have levels of THC in your blood for up to 25 days.
* **Hair strand test**
  The most sensitive of all the tests to low levels, and you can get a positive test up to 90 days after taking it.[32]

16

WHAT IS THE BEST CANNABIS POLICY?

IF YOU'VE READ this far, you're probably in favour of cannabis reform. It seems public opinion in the UK is now broadly pro, too. An April 2021 YouGov poll found that 52 per cent would support the legalisation of cannabis, against 32 per cent who would oppose it (15 per cent said they didn't know).[1]

Those numbers suggest that if we had a different political system in the UK, we might already have legalised. In US states, it was the voters who changed the law directly, because the US system meant this issue could be included on election ballots.

But is legalisation the right route? Or is decriminalisation, no longer penalising people for personal possession of cannabis, better? If cannabis is legalised, should it be a free market or should there be some state control? The advantage for the UK of coming late to reform is that we can learn from the territories that have already reformed such as parts of the US and Canada.

So what might an evidence-based cannabis policy look like? In 2015, Drug Science conducted a landmark piece of research to find this out. Our study cut across all political and ethical beliefs, starting from first principles rather than a stated position. This kind of evidence-based approach is important for big social issues: the experience of Covid has taught us that people only obey policies they believe do more good than harm.

We did this research with the Research Council of Norway. Traditionally, Norway had a very punitive approach to drugs and drug users. But although it's one of the richest countries in the world, it also had one of the highest death rates for drug users, almost three times the European Union average. This was thought to be due to users of drugs such as heroin avoiding treatment for fear of being penalised.[2]

The Norwegians wanted to work out how to minimise the harms of cannabis while maximising the benefits.

Drugs policies impact so many aspects of all our lives, and that makes the possible effects hard to unravel, or predict. We've covered health and mental health in detail in the book as well as road safety, but we've barely touched on health costs, crime, policing and social injustice. For example, in the UK, the black and Caribbean community have suffered disproportionately from the law on cannabis and its enforcement in the past 50 years. In the US and Canada, there has been a lot of useful debate on how to change the law in a way that prevents discrimination, indeed to make reparations for past injustices.[3]

If we did choose some form of regulated market for cannabis, there is the prospect of tax income, which should appeal to our Covid-impoverished governments. In Colorado, for example, cannabis has an extra 15 per cent tax that's earmarked specifically to build schools.[4]

For the study, we used the same type of decision theory, multi-criteria decision analysis (MCDA), as we did for drug harms (see Chapter Four).

MCDA allows you to reduce each policy and outcome into a set of simpler judgements, giving numerical values so you can measure them against each other.[5]

We began by bringing together a group of experts from various disciplines across drug harms, addiction, criminology and drug policy, including an economist, pharmacologist, policy analyst, vet, criminologist, epidemiologist, social scientists, psychiatrist, a pharmacist, an ex-Home Office drug department civil servant, an ex-chief constable of police, an addiction sociologist and a neuropsychopharmacologist. It was led by Ole Rogeberg, a professor of economics from Oslo, and coordinated by the same decision theory expert as before, Larry Phillips.

Next, we defined our four main options for regulatory regimes:

1 Absolute prohibition. This is what we have in the UK, and what most of the world still has too.
2 Prohibition with decriminalisation of possession/use. This is what happens in the Netherlands.

3 Legally regulated markets under strict state control, as in Uruguay. There, the president introduced state control with cannabis sold in pharmacies.

4 Legal commercial 'free' markets, as in Colorado and Canada.

We went on to define the potential effects on all aspects of society.

Then we worded all the effects as benefits, in order to be able to compare them. For example, harm to users became how much a policy would reduce harms to users.

Finally, we sorted all the benefits into seven categories: health, social, political, public, crime, economic and cost – see table below.

TOPIC	IMPACT
Health	Reduces user medical harms
	Reduces harms to third parties, e.g. second-hand smoke
	Shifts use to lower-harm products, e.g. reduces use of spice
	Encourages treatment for substance use
	Improves product quality, e.g. dosage and purity
Social	Promotes drug education
	Enables medical use
	Promotes/supports research
	Protects human rights
	Promotes individual liberty
	Improves community cohesion
	Promotes family cohesion

CANNABIS

Political	Supports international development/security
	Reduces drug industry influence – less lobbying
Public	Promotes social and personal well-being
	Protects the young
	Protects vulnerable groups
	Respects religious/cultural values
Crime	Reduces criminalisation of users
	Reduces acquisitive crime to finance use
	Reduces violent crime due to illegal markets
	Prevents corporate crime, e.g. money laundering
	Prevents criminal industry
Economic	Generates state revenue
	Reduces economic costs, e.g. on health policy budgets
Cost	Low policy introduction costs
	Low policy maintenance costs

At a second conference, we quantified the impact of the policies on each of the 27 criteria. We did this by putting them on a 0 to 100 scale, with the best given 100 and the rest rated proportionately.

We used data from different countries where we had it, and expert opinions and debate where that was inadequate or missing.

Finally, each of the 27 criteria were weighted according to how important we thought they were, and all the data was fed into the computer.

*　　*　　*

And the winner was? State control. It came out best overall; in particular it turned out to be best for social and health benefits.[6,7]

As you can see from the graph below, state control was followed by free market, then decriminalisation. Absolute prohibition came last.

In fact it was really surprising quite how poorly absolute prohibition scored, much worse than all the other policies for almost all the criteria.

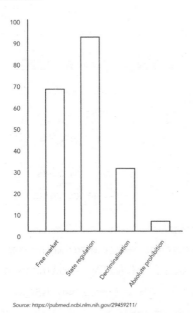

Source: https://pubmed.ncbi.nlm.nih.gov/29459211/

Fig. 13 Comparative benefits of four different policy options

We also looked at what would happen if we prioritised different criteria. But we found that state control remained the best option for almost all changes.

This may sound unappealing to you, living in a Western democracy where the general trend is towards the free market and liberalisation.

But if you think about it, we are comfortable with the NHS being under state control, as health is such an important aspect of our lives. You could argue the same approach should be taken with cannabis, too.

I've presented this study data at the All Parliamentary Group for Drug Policy Reform. And I've offered this process to various politicians in the UK. Needless to say, so far I haven't been taken up on it.

FOLLOWING THE SCIENCE?

The government could continue to ignore the evidence on cannabis policies. But I really hope politicians will take a lesson from Covid-19, when the government repeatedly declared its willingness to 'follow the science' to make policy decisions.

In fact, it was public opinion rather than science, stirred up by the media, that forced the government to change the legal status of medical cannabis. They had to react to stories in the newspapers of very sick children having their medicines taken away. Even those who'd been steadfastly anti-cannabis, such as Theresa May, couldn't ignore the spectre of dying children.

However, as you already know, there are still very few prescriptions for cannabis medicines being provided on the NHS.

This is similar to the situation in Canada ten years ago. Medicinal cannabis was legal but, with many doctors who were suspicious of cannabis and who had not been educated about it, very few patients gained access.

It's not clear how this will be solved in the UK: education in cannabis for medical practitioners will help, as will more research and building up evidence. The Canadian solution was to legalise recreational cannabis, in order to open up access, quickly and cheaply, to people who would benefit from using it for medicinal purposes.

This seems unlikely to happen in the UK. Here, cannabis legalisation is still seen as a vote-losing hot potato that neither of the two main UK parties will touch. Unsurprisingly for a former prosecutor, Keir Starmer has now said he is in favour of keeping the status quo.

More hopefully, the London mayor Sadiq Khan recently came out in favour of decriminalisation in London after a 2019 Volteface survey of Londoners showed that 63 per cent agreed with the legalisation of cannabis.[8]

But in response, Boris Johnson's then press secretary, Allegra Stratton, said this: 'Policy on controlled drugs is a matter for UK government and there are no plans to devolve this responsibility.

'The prime minister has spoken about this on many occasions – illicit drugs destroy lives and he has absolutely no intention of legalising cannabis, which is a harmful substance.'

Johnson is a victim of 50 years of political brainwashing that cannabis is a dangerous substance. He can't see that the drug laws are not fit for purpose. He can only see: drugs are bad, so stop all drugs. This is what drug reformers come up against time and again: fixed views, underpinned by emotion and moral judgements and driven by political expediency.

In the UK, there are politicians from both parties who can see that it is just not sustainable to carry on with present policies, such as Conservative MP Crispin Blunt, who has said he thinks cannabis will be legalised in five years: 'The product will never be safe, just like drinking alcohol or smoking tobacco . . . risks can be highlighted, age limits can be introduced, so the consumer can be sure what they are buying and what its potency is.'[9]

The Labour MP Jeff Smith is Chair of the All-Party Parliamentary Group on Drug Policy Reform – a cross-party group of MPs and peers campaigning for evidence-based drug policy. He is also a trustee of Drug Science.

He has written: 'A legal, regulated cannabis market would protect us from harm in the same way as alcohol licensing prevents us from purchasing 80% proof moonshine or toxic tobacco.

'Last year 2.1 million people used cannabis, bought from dealers with little regard for the safety of their product.

'Regulation could prevent particularly harmful forms of cannabis, such as the high-strength skunk proliferating in the black market, from reaching our streets and bring in an estimated £1bn per year in tax revenue.

'That money could be spent on healthcare or policing, instead of going into the pockets of criminal gangs.'[10]

One party, the Liberal Democrats, have taken a rational stance. They set up a special committee – I was a member – to examine cannabis policies. We concluded that a regulated market was the way forward, then the Liberal Democrats adopted it as a manifesto commitment in 2016. However, they haven't had any opportunity to take action on this.

In addition to the two major parties, there are some powerful media, business and religious interests in favour of keeping cannabis illegal. In 2020, New Zealand had a referendum on changing the law on cannabis. It looked as if this was a done deal. But the reformers narrowly lost, 50.7 per cent to 48.4 per cent, beaten by a very sophisticated anti-cannabis media campaign whose funding it is not possible to uncover.[11]

DE FACTO DECRIMINALISATION

In the UK, what seems more likely than a change in the law is de facto decriminalisation.

A growing number of police authorities are moving away from arresting people caught with small amounts of drugs, recognising that this takes up too much police time and resources. Being prosecuted doesn't make it less likely that someone will reoffend. In fact, being arrested appears to make it more likely that people will reoffend, because criminal records are so damaging to people's life chances.

There are now five police forces in England and Wales that have stopped criminalising people caught with small amounts of drugs. Instead, they are setting up diversion schemes, where instead of being arrested people are given harm reduction and treatment advice.[12]

In fact, contrary to what Boris Johnson believes, Sadiq Khan can make police policy. Being mayor makes him the equivalent of the Police and Crime Commissioner in London, responsible for London's Police and Crime Plan. So in theory, he can decide how to treat people who are caught in possession of cannabis.[13, 14]

As we saw in Chapter 14, Arfon Jones, the ex-Police and

Crime Commissioner for North Wales, has been outspoken about the need for more rational cannabis policies in prison. He recently wrote to the Home Office, asking for UK cannabis patients not to be criminalised. He said: 'In my policing career I have never met anyone who has caused violence through cannabis, as opposed to the hundreds of violence cases I have seen related to alcohol, which is a legal substance.

'It is unfair that a conviction for minor cannabis possession can blight a person's future career as that's what happens when people go through the criminal justice system.[15]

LEARNING FROM EXPERIENCE

Evidence coming from countries that have decriminalised or legalised doesn't seem to have brought up any big concerns.

But we knew that from the Dutch experience of more than 30 years. Decriminalising the possession of small amounts of cannabis and creating the coffee shops was designed to separate cannabis from the illicit markets for more harmful drugs such as heroin and cocaine. It worked; Dutch youth use of these harder drugs went down, while at the same time in the UK it went up. And cannabis use was lower than in the UK.[16]

And they don't have a problem with spice in the Netherlands, as we do in the UK.

Going back to Norway, the government decided not to go down the state control route. There was recently a vote as to whether cannabis – and other drugs in small amounts – should be decriminalised. But it wasn't passed. The socialist party blocked it – as the Labour Party appears to be doing in the UK.

CAN WE STOP OUR KIDS TAKING CANNABIS?

One question that comes up again and again – rightly – is that if cannabis is legalised, will more underage young people take it?

One of the major benefits of state control is that it gives young people much greater protection than the free market.

For example, in Canada, teenage use has gone down since its legalisation for recreational use in 2018. Statistics from Canada's National Cannabis Survey has shown that the percentage of people aged 15 to 17 who use cannabis has fallen from 20 per cent to 10 per cent.[17]

Teens in the US are less likely to use cannabis once it's been legalised for recreational use too. It's thought that this is because it's more difficult to buy from licensed shops, as proof of age is needed, than it is from dealers.[18, 19]

Other data have suggested some increased use in teenagers and the issue is still being debated.[20, 21]

Only time will tell what the final situation will be. But even if use by teens does increase, the elimination of the illicit market would be a major gain, in the same way that the repeal of alcohol prohibition in the 1930s helped curb organised crime.[22]

State control can include limiting the sale of high-strength skunk without any CBD, proper labelling of THC content and limiting sales to minors. And it would allow all young people to be properly educated about cannabis, as well as enable the monitoring of any marketing.

I was concerned about the originally very libertarian approach when cannabis was first legalised in Colorado, which in effect allowed the sale of any amount of cannabis in any form. I was worried that naive users, especially adolescents, might eat a few gummy bears without understanding the dose of THC in them and so accidentally become very stoned. Remember the group of schoolgirls in London (in Chapter Seven) who were sent to hospital having eaten gummy bears containing THC? If cannabis was under state control, they would (hopefully) not have been able to buy them.

Some similar instances, of underage people taking THC, happened in Colorado and in 2014, the state enacted new laws to ban packaging that appeals to children. Packages must now be clearly marked to keep them out of the reach of children, must show their serving size (one serving is 10mg THC), and they can no longer be described on the packaging as candy.[23]

Colorado's experiment with a totally free market did reveal some problems, as our research predicted. The best result, if the politicians can hold their nerve, is state control. This would bring the benefits of age barriers, clear labelling, strength thresholds and dose-related taxation. Perhaps the latter – along with public opinion being in favour of legalisation – will finally convince politicians that prohibition can be overturned.

CONCLUSION

MOST PEOPLE AREN'T STUPID and would not take something that's as dangerous as cannabis is claimed to be. Despite the fact that politicians and the media have been telling us how dangerous cannabis is for 100 years, it's still the most used illicit drug in the world.

People know cannabis is less likely to kill you than either alcohol or tobacco. It's not 100 per cent safe, but what drug or medicine is? So most sensible people conclude that cannabis isn't harmful enough to be banned – and certainly not to be denied to medical patients.

Cannabis isn't a miracle medicine or a wonder drug. But most of its unwanted effects come from the fact that most of it is being bought on the illicit market, and what's available is mainly skunk. More access to safer and good-quality cannabis with a balanced mixture of CBD and THC, plus medical advice, could improve quality of life for hundreds of thousands of people.

If you are considering using cannabis or do use it, do read up on the risks of dependence and mental health issues. Avoid skunk, if you can. For medical cannabis, try the Drug Science Twenty21 initiative. And be aware that if you are using cannabis recreationally you're breaking the law. If your teenager is using it, have a serious conversation with them about it, and give them this book to read.

Most doctors say that more research is needed for medicinal cannabis, and I agree with them. But we know cannabis is different from other medicines because of its long history of use as a plant medicine. We know it's comparatively safe.

We do have some good evidence, but it's not the kind currently required by regulators, which is randomised controlled trials. However, we do have a growing body of real-world evidence, which can ultimately provide good-quality measures of outcomes in real patients. In the meantime, we should give patients access to medical cannabis as a compassionate act, while monitoring the benefits and any adverse effects using standardised methods, as we are doing in Twenty21.

We also need to educate doctors that their fears about medical cannabis leading to psychosis and dependence are unfounded. These are largely a product of street cannabis, high-THC skunk. Doctors should be taught about medical cannabis in the early years of medical school, with an emphasis on the remarkable features of the endocannabinoid system with its myriad nurturing and stabilising actions on the brain and immune systems.

As has been shown in Canada, legalisation may be the way to open up access to everyone who needs cannabis for medical reasons. This would help those millions of patients whose access is currently being blocked by a combination of medical ignorance, legal hurdles, doctors' resistance and the refusal of the NHS to accept its cost-effectiveness.

When looked at purely on the evidence, state control is the best option. A regulated market would ensure that products are safe and contain a good balance of ingredients.

CONCLUSION

Writing this book, bringing together all my work and others' work, all the facts about cannabis in one place, has made me even more convinced that we need to change our policy, and do it intentionally, for everyone's benefit.

THINKING OF USING CANNABIS? HERE'S WHAT TO CONSIDER:

* If cannabis is not legal where you live, are you willing to take the risk of breaking the law? Does what you're using it for – for example pain relief – make it worth that risk? Is there a way to get cannabis legally for your condition?
* Do you have a family history of psychosis? Although cannabis isn't thought to cause schizophrenia, it may precipitate it in some people who have a genetic tendency.
* How is your mental health? Cannabis can produce unpleasant mental short-term effects, such as paranoia and psychosis.
* Do you have a propensity to dependence? Are you at a difficult time of life? It's thought that 8 to 10 per cent of people who use cannabis become dependent.
* Do you have chronic lung disease? It's not a good idea to smoke or vape, although you can take it orally.
* Do you have heart disease and/or high blood pressure? Cannabis can (rarely) bring on a heart attack.
* Are you under 18? It's not advisable to use cannabis as a teen, in particular under the age of 15.
* If you have any medical issues or concerns or are taking any medication, you should always speak to your GP first.

APPENDIX: CUDIT-R QUESTIONNAIRE

The Cannabis Use Disorder Identification Test - Revised (CUDIT-R)

Have you used any cannabis over the past six months?   YES / NO

If YES, please answer the following questions about your cannabis use. Circle the response that is most correct for you in relation to your cannabis use *over the past six months*

1. How often do you use cannabis?

Never	Monthly or less	2-4 times a month	2-3 times a week	4 or more times a week
0	1	2	3	4

2. How many hours were you "stoned" on a typical day when you had been using cannabis?

Less than 1	1 or 2	3 or 4	5 or 6	7 or more
0	1	2	3	4

3. How often during the past 6 months did you find that you were not able to stop using cannabis once you had started?

Never	Less than monthly	Monthly	Weekly	Daily or almost daily
0	1	2	3	4

4. How often during the past 6 months did you fail to do what was normally expected from you because of using cannabis?

Never	Less than monthly	Monthly	Weekly	Daily or almost daily
0	1	2	3	4

5. How often in the past 6 months have you devoted a great deal of your time to getting, using, or recovering from cannabis?

Never	Less than monthly	Monthly	Weekly	Daily or almost daily
0	1	2	3	4

6. How often in the past 6 months have you had a problem with your memory or concentration after using cannabis?

Never	Less than monthly	Monthly	Weekly	Daily or almost daily
0	1	2	3	4

7. How often do you use cannabis in situations that could be physically hazardous, such as driving, operating machinery, or caring for children:

Never	Less than monthly	Monthly	Weekly	Daily or almost daily
0	1	2	3	4

8. Have you ever thought about cutting down, or stopping, your use of cannabis?

Never	Yes, but not in the past 6 months	Yes, during the past 6 months
0	2	4

This scale is in the public domain and is free to use with appropriate citation:

Adamson SJ, Kay-Lambkin FJ, Baker AL, Lewin TJ, Thornton L, Kelly BJ, and Sellman JD. (2010). An Improved Brief Measure of Cannabis Misuse: The Cannabis Use Disorders Identification Test – Revised (CUDIT-R). *Drug and Alcohol Dependence* 110:137-143.

NOTES

INTRODUCTION

1 https://www.ncsl.org/research/health/state-medical-marijuana-laws.aspx
2 https://www.gov.uk/government/publications/cannabis-based-products-for-medicinal-use-in-humans-cbpms
3 https://journals.sagepub.com/doi/full/10.1177/2050324520974487

CHAPTER ONE

1 https://hashmuseum.com/en/collection/reefer-madness
2 https://en.wikipedia.org/wiki/Harry_J._Anslinger
3 https://www.sciencemag.org/news/2019/06/oldest-evidence-marijuana-use-discovered-2500-year-old-cemetery-peaks-western-china
4 https://www.researchgate.net/publication/316545890_History_of_medical_cannabis
5 https://www.visualcapitalist.com/history-medical-cannabis-shown-one-giant-map
6 https://publications.parliament.uk/pa/ld199798/ldselect/ldsctech/151/15103.htm

CHAPTER TWO

1 https://www.newscientist.com/article/mg23831760-400-high-times-the-victorian-doctor-who-promoted-medical-marijuana/#ixzz6t8HEKlld
2 https://acnr.co.uk/2019/07/john-russell-reynolds
3 https://pubmed.ncbi.nlm.nih.gov/4557840
4 https://www.tni.org/files/download/rise_and_decline_ch1.pdf
5 https://vinepair.com/articles/alcohol-companies-that-dont-like-weed
6 https://drugscience.org.uk/better-late-than-never-after-82-years-the-who-reviews-cannabis

CHAPTER THREE

1 https://core.ac.uk/download/pdf/191532235.pdf
2 http://www.police-foundation.org.uk/publication/inquiry-into-drugs-and-the-law
3 http://news.bbc.co.uk/1/hi/uk_politics/697937.stm
4 https://www.legislation.gov.uk/uksi/2003/3201/made
5 https://assets.publishing.service.gov.uk/government/uploads/system/uploads/attachment_data/file/119126/cannabis-class-misuse-drugs-act.pdf
6 https://core.ac.uk/download/pdf/191532235.pdf
7 http://eprints.lse.ac.uk/100751/1/TheColourOfInjustice.pdf
8 https://core.ac.uk/download/pdf/191532235.pdf
9 https://www.thelancet.com/journals/lancet/article/PIIS0140 6736(07)60464-4/fulltext
10 https://assets.publishing.service.gov.uk/government/uploads/system/uploads/attachment_data/file/119174/acmd-cannabis-report-2008.pdf
11 https://hansard.parliament.uk/Commons/2008-05-07/debates/08050765000005/Cannabis?highlight=cannabis#contribution-08050765000177
12 https://www.crimeandjustice.org.uk/sites/crimeandjustice.org.uk/files/Estimating%20drug%20harms.pdf
13 https://www.theguardian.com/politics/2009/nov/01/david-nutt-alan-johnstone-drugs

CHAPTER FOUR

1 https://www.thelancet.com/journals/lancet/article/PIIS0140
6736(10)61462-6/fulltext
2 https://www.thelancet.com/journals/lancet/article/PIIS0140
6736(07)60464-4/fulltext#figures
3 https://www.ncbi.nlm.nih.gov/pmc/articles/PMC7846668
4 https://www.sciencedaily.com/releases/2018/01/180109214939.htm

CHAPTER FIVE

1 https://www.theguardian.com/society/shortcuts/2015/aug/31/
dabbing-cannabis-crack-concentrated-oil
2 https://www.ncbi.nlm.nih.gov/pmc/articles/PMC4851925
3 https://www.england.nhs.uk/medicines-2/support-for-prescribers/
cannabis-based-products-for-medicinal-use/cannabis-based-
products-for-medicinal-use-frequently-asked-questions/#what-are-
synthetic-cannabinoids-and-are-they-included-in-the-re-scheduling
4 https://www.zmescience.com/medicine/skunk-marijuana-smell
-043222
5 https://analyticalsciencejournals.onlinelibrary.wiley.com/doi/abs/
10.1002/dta.2368
6 https://assets.publishing.service.gov.uk/government/uploads/
system/uploads/attachment_data/file/882953/Review_of_Drugs_
Evidence_Pack.pdf
7 https://www.theguardian.com/society/2017/mar/25/trafficked-
enslaved-teenagers-tending-uk-cannabis-farms-vietnamese
8 https://www.theguardian.com/news/2019/jul/26/vietnamese-
cannabis-farms-children-enslaved

CHAPTER SIX

1 https://en.wikipedia.org/wiki/Rimonabant

CHAPTER SEVEN

1 https://www.semanticscholar.org/paper/Cannabis-and-schizophrenia%3A-model-projections-of-of-Hickman-Vickerman/ca53a07ad7c7f51690c92ff9472d4adf35003df3/figure/1
2 https://jamanetwork.com/journals/jamanetworkopen/fullarticle/2716990
3 https://www.sciencedaily.com/releases/2018/12/181204131115.htm
4 https://pubmed.ncbi.nlm.nih.gov/28531767
5 https://www.ncbi.nlm.nih.gov/pmc/articles/PMC5890870
6 https://www.dailymail.co.uk/news/article-9533379/Three-boys-girl-taken-hospital-eating-jelly-sweets-contained-cannabis.html
7 https://www.ons.gov.uk/peoplepopulationandcommunity/birthsdeathsandmarriages/deaths/datasets/deathsrelatedtodrugpoisoningbyselectedsubstances
8 https://pubs.niaaa.nih.gov/publications/arh27-1/39-51.pdf
9 https://tinyurl.com/46dxhuuw
10 https://pubmed.ncbi.nlm.nih.gov/32735782
11 https://druglibrary.org/schaffer/hemp/history/first12000/12.htm
12 https://pubmed.ncbi.nlm.nih.gov/8797240
13 https://pubmed.ncbi.nlm.nih.gov/31013455
14 https://www.sciencedaily.com/releases/2018/07/180717094747.htm
15 https://www.med.upenn.edu/cbti/assets/user-content/documents/s11920-017-0775-9.pdf
16 http://www.bandolier.org.uk/bandopubs/cannfly/cannfly.html

CHAPTER EIGHT

1 https://www.rollingstone.com/feature/cannabis-legalization-states-map-831885
2 https://www.fool.com/investing/stock-market/market-sectors/healthcare/marijuana-stocks/marijuana-legalization
3 https://assets.publishing.service.gov.uk/government/uploads/system/uploads/attachment_data/file/939090/OFFICIAL__Published_version_-_ACMD_CBPMs_report_27_November_2020_FINAL.pdf

4 https://thecmcuk.org/news/million-uk-adults-self-medicating-with
-illicit-cannabis

5 https://www.thetimes.co.uk/article/medicinal-cannabis-hannah-
deacon-interview-alfie-dingley-cs2q0x6fv

6 https://time.com/pot-kids

7 https://www.nytimes.com/2020/04/09/us/charlotte-figi-dead.html

8 https://blogs.bmj.com/bmj/2020/08/28/hannah-deacon-patients-
still-do-not-have-access-to-medical-cannabis

9 https://www.cannabis-med.org/index.php?tpl=page&id=
235&lng=en

10 https://www.ukmccs.org/information/patient-story/hannah-
deacon-fighting-for-access-to-medical-cannabis

11 https://metro.co.uk/2020/07/13/cannabis-treatment-saved-sons-life
-12983294

12 https://www.independent.co.uk/news/health/billy-caldwell-
medical-cannabis-epilepsy-marijuana-nhs-customs-confiscate-
a8393401.html

13 https://www.dailymail.co.uk/health/article-7367037/Alfie-Dingley-
one-year-changed-law-medicinal-cannabis.html

14 https://assets.publishing.service.gov.uk/government/uploads/
system/uploads/attachment_data/file/722010/CMO_Report_
Cannabis_Products_Web_Accessible.pdf

CHAPTER NINE

1 https://www.sentencingcouncil.org.uk/wp-content/uploads/Drug-
offences-definitive-guideline-Web.pdf

2 https://www.theguardian.com/society/2019/sep/20/mother-
sells-house-buy-daughter-medical-cannabis

3 https://pubmed.ncbi.nlm.nih.gov/33970291

4 https://www.bbc.co.uk/news/health-50351868

5 https://www.nice.org.uk/guidance/ng144/chapter recommendations-
for-research#key-recommendations-for-research

6 https://journals.sagepub.com/doi/full/10.1177/2050324520974487

7 https://www.dailymail.co.uk/health/article-7367037/Alfie-
Dingley-one-year-changed-law-medicinal-cannabis.html

8 https://journals.sagepub.com/doi/full/10.1177/0269881120986393

9 https://www.nice.org.uk/guidance/ng144/chapter/recommendations-for-research#key-recommendations-for-research
10 https://journals.sagepub.com/doi/full/10.1177/2050324520974487

CHAPTER TEN

1 https://www.bmj.com/content/365/bmj.l1141
2 https://www.ncbi.nlm.nih.gov/pmc/articles/PMC4311234/figure/f2
3 https://www.nature.com/articles/srep08126
4 https://www.food.gov.uk/safety-hygiene/cannabidiol-cbd
5 https://www.health.harvard.edu/blog/cannabidiol-cbd-what-we-know-and-what-we-dont-2018082414476
6 https://onlinelibrary.wiley.com/doi/10.1111/epi.13852
7 https://www.cdprg.co.uk/blog/2019/7/4/its-a-myth-that-we-can-arrest-our-way-out-of-drug-problems-mike-barton-speech-at-cdprg-launch.
8 https://irp-cdn.multiscreensite.com/51b75a3b/files/uploaded/Report%20%7C%20CBD%20in%20the%20UK%20%20%20Exec%20Summary.pdf
9 https://pubmed.ncbi.nlm.nih.gov/33970291

CHAPTER ELEVEN

1 https://www.sciencedirect.com/science/article/abs/pii/S0028390821001404
2 https://www.ncbi.nlm.nih.gov/pmc/articles/PMC6135562
3 https://www.drugscience.org.uk/medical-cannabis-compared-to-common-pain-medications
4 https://theconversation.com/1-in-10-women-with-endometriosis-report-using-cannabis-to-ease-their-pain-126516
5 https://journals.sagepub.com/doi/full/10.1177/2050324520974487
6 https://www.ncbi.nlm.nih.gov/pmc/articles/PMC6235654

7 https://acnr.co.uk/2021/03/seizures-associated-with-tuberous-sclerosis-complex/

8 https://epilepsysociety.org.uk/living-epilepsy/sudep

9 https://www.dovepress.com/cannabinoids-in-the-treatment-of-epilepsy-current-status-and-future-pr-peer-reviewed-fulltext-article-NDT

10 https://mstrust.org.uk/news/sativex-cost-effective-final-nice-guideline-cannabis-based-medicines

11 https://academic.oup.com/crohnscolitis360/article/2/2/otaa015/5821009

12 https://www.mcmasteroptimalaging.org/blog/detail/blog/2020/02/18/does-cannabis-offer-new-hope-for-folks-with-crohn-s-disease-and-ulcerative-colitis

13 https://pubmed.ncbi.nlm.nih.gov/24614667

14 https://www.ajmc.com/view/cannabis-shown-to-relieve-parkinson-disease-symptoms

15 https://www.mdpi.com/1420-3049/24/5/918

16 https://pubmed.ncbi.nlm.nih.gov/18181976

17 https://pubmed.ncbi.nlm.nih.gov/30993303

18 http://scholar.google.co.uk/scholar_url?url=https://www.mdpi.com/2077-0383/8/6/807/pdf&hl=en&sa=X&ei=YLwOYJreO4fCmgH5tb6ABg&scisig=AAGBfm2vIQhWNcnr2sU8L7ZN3PHrL-Lnow&nossl=1&oi=scholarr

19 https://www.karger.com/Article/FullText/496355

20 https://neuro.psychiatryonline.org/doi/full/10.1176/appi.neuropsych.16110310

21 https://www.nature.com/articles/s41598-018-37570-y

22 https://www.ncbi.nlm.nih.gov/pmc/articles/PMC6834767

23 https://bmjopen.bmj.com/content/10/12/e043400

24 https://pubmed.ncbi.nlm.nih.gov/32735782

25 https://pubmed.ncbi.nlm.nih.gov/28194850

26 https://pubmed.ncbi.nlm.nih.gov/31315244

27 https://www.forbes.com/sites/emilyearlenbaugh/2020/07/01/cbd-for-dogs-new-research-backs-canine-cannabis-use-for-osteoarthritis/?sh=58eddc9a40ac

28 https://pubmed.ncbi.nlm.nih.gov/16282192

29 https://www.thecannachronicles.com/the-english-patient-1840

30 https://www.ncbi.nlm.nih.gov/pmc/articles/PMC7388834

31 https://link.springer.com/article/10.1007/s11920-017-0775-9

32 https://pubmed.ncbi.nlm.nih.gov/32949837

33 https://pubmed.ncbi.nlm.nih.gov/21307846
34 https://pubmed.ncbi.nlm.nih.gov/28553229
35 https://pubmed.ncbi.nlm.nih.gov/31109198
36 https://www.ncbi.nlm.nih.gov/pmc/articles/PMC6326553
37 https://www.mdpi.com/1648-9144/55/9/525
38 https://pubmed.ncbi.nlm.nih.gov/32469819
39 https://pubmed.ncbi.nlm.nih.gov/24987795
40 https://www.liebertpub.com/doi/pdf/10.1089/can.2020.0102
41 https://clinicaltrials.gov/ct2/show/study/NCT04504877
42 https://www.ncbi.nlm.nih.gov/pmc/articles/PMC4882033
43 https://pubmed.ncbi.nlm.nih.gov/28576350
44 https://pubmed.ncbi.nlm.nih.gov/31994476
45 https://pubmed.ncbi.nlm.nih.gov/19924114
46 https://www.ncbi.nlm.nih.gov/pmc/articles/PMC3316151
47 https://pubmed.ncbi.nlm.nih.gov/29241357
48 https://pubmed.ncbi.nlm.nih.gov/31994476
49 https://pubmed.ncbi.nlm.nih.gov/32526965
50 https://pubmed.ncbi.nlm.nih.gov/32758396
51 https://pubmed.ncbi.nlm.nih.gov/31715263
52 https://pubmed.ncbi.nlm.nih.gov/23795762
53 https://jamanetwork.com/journals/jamapsychiatry/fullarticle/
    2723657
54 https://thecmcuk.org/news/million-uk-adults-self-medicating-with
    -illicit-cannabis
55 https://pubmed.ncbi.nlm.nih.gov/33228239
56 https://www.nature.com/articles/d41586-018-05261-3
57 https://pubmed.ncbi.nlm.nih.gov/29482741
58 https://pubmed.ncbi.nlm.nih.gov/30627539
59 https://www.sciencedirect.com/science/article/abs/pii/
    S1043661820316108
60 https://pubmed.ncbi.nlm.nih.gov/32575540
61 https://www.ncbi.nlm.nih.gov/pmc/articles/PMC7554803
62 https://www.spinalcord.com/blog/how-medical-cannabis-cbd-can-
    help-people-with-spinal-cord-injuries
63 https://pubmed.ncbi.nlm.nih.gov/1839644
64 https://huntingtonsdiseasenews.com/2019/02/21/fda-awards-
    orphan-drug-designation-to-mmj-002-for-huntingtons
65 https://www.healtheuropa.eu/motor-neuron-disease-symptoms-
    cannabis-sativa-plant/89408
66 https://mndresearch.blog/2019/01/24/cannabis-based-products-for-

medicinal-use
67 https://www.tandfonline.com/doi/abs/10.3109/
00952990.2010.500438
68 https://academic.oup.com/aje/article/174/8/929/155851
69 https://onlinelibrary.wiley.com/doi/abs/10.1002/oby.20973
70 https://pubmed.ncbi.nlm.nih.gov/32360935
71 https://finfeed.com/small-caps/biotech/can-bda-medical-cannabis-
product-treat-long-covid-symptoms

CHAPTER TWELVE

1 https://www.who.int/substance_abuse/publications/msbcannabis.pdf
2 https://www.gov.uk/government/statistics/substance-misuse-
treatment-for-adults-statistics-2019-to-2020/adult-substance-
misuse-treatment-statistics-2019-to-2020-report
3 https://www.drugabuse.gov/publications/research-reports/
marijuana/marijuana-addictive
4 https://pubmed.ncbi.nlm.nih.gov/17236542
5 https://onlinelibrary.wiley.com/doi/abs/10.1111/adb.12116
6 https://jamanetwork.com/journals/jamapsychiatry/article-abstract/
491465
7 https://www.healthline.com/health/marijuana
-withdrawal#symptoms
8 https://pubmed.ncbi.nlm.nih.gov/29382407
9 https://static1.squarespace.com/static/5f1ebab9df1a5a6c
6f4a9fd0/t/608ada0a9b3f2c05c53b74d9/1619712526736/
Left+Behind+.pdf
10 https://www.thelancet.com/journals/lancet/article/
PIIS0140-6736(10)61462-6/fulltext
11 https://pubmed.ncbi.nlm.nih.gov/21145178
12 https://www.researchgate.net/profile/Lynn-Warner-2/publication/
232545123_Comparative_Epidemiology_of_Dependence_on_
Tobacco_Alcohol_Controlled_Substances_and_Inhalants_Basic_
Findings_From_the_National_Comorbidity_Survey/links/0fcfd-
5124debe9ee41000000/Comparative-Epidemiology-of-Dependence
-on-Tobacco-Alcohol-Controlled-Substances-and-Inhalants-Basic-
Findings-From-the-National-Comorbidity-Survey.pdf

13 https://pubmed.ncbi.nlm.nih.gov/32735782
14 https://www.thelancet.com/journals/lanpsy/article/PIIS2215
0366(20)30290-X/fulltext
15 https://pubmed.ncbi.nlm.nih.gov/31305874
16 https://pubmed.ncbi.nlm.nih.gov/27149547
17 https://marijuana-anonymous.org
18 https://pubmed.ncbi.nlm.nih.gov/27149547
19 https://gspp.berkeley.edu/assets/uploads/research/pdf/BJP2001_
CannabisRegimes.pdf
20 https://www.ncbi.nlm.nih.gov/pmc/articles/PMC5569620
21 https://www.ncbi.nlm.nih.gov/pmc/articles/PMC4311234

CHAPTER THIRTEEN

1 https://www.theguardian.com/politics/2009/nov/02/drug-policy-alan-johnson-nutt
2 https://hansard.parliament.uk/Commons/2009-11-02/debates/
2549ce62-a1ef-4a5f-8772-f6bf1b36adec/CommonsChamber
3 https://pubmed.ncbi.nlm.nih.gov/26246443
4 https://www.pnas.org/content/95/14/8268
5 https://www.hindawi.com/journals/cjgh/2018/9430953
6 Simmington, Rupert: *Cannabis: A Study of its History, Prohibition and Use*, independently published, 2019
7 Ibid.
8 https://pubmed.ncbi.nlm.nih.gov/2892048
9 https://www.nature.com/articles/1300496
10 https://pubmed.ncbi.nlm.nih.gov/26781550
11 https://pubmed.ncbi.nlm.nih.gov/16399153
12 https://www.ias.org.uk/uploads/pdf/Factsheets/FS%20price%20
032017.pdf
13 Ibid.
14 https://citeseerx.ist.psu.edu/viewdoc/download?doi=
10.1.1.525.5088&rep=rep1&type=pdf
15 https://pubmed.ncbi.nlm.nih.gov/19560900
16 https://pubmed.ncbi.nlm.nih.gov/29117908
17 https://pubmed.ncbi.nlm.nih.gov/25471704
18 https://pubmed.ncbi.nlm.nih.gov/14992976

NOTES

19 https://www.nature.com/articles/s41593-018-0206-1

20 https://www.thelancet.com/journals/lanpsy/article/PIIS2215
0366(14)00117-5/fulltext

21 https://pubmed.ncbi.nlm.nih.gov/26777297

22 https://realmofcaring.org/wp-content/uploads/2020/03/
Cannabidiol-CBD-as-an-Adjunctive-Therapy-in-Schizophrenia-A-
Multicenter-Randomized-Controlled-Trial.pdf

23 https://pubmed.ncbi.nlm.nih.gov/22832859

24 https://jamanetwork.com/journals/jamanetworkopen/fullarticle/
2772562

25 https://www.ons.gov.uk/peoplepopulationandcommunity/crime-
andjustice/adhocs/12606proportionandfrequencyof16to17year
oldsreportinguseofcannabisinthelastyearforyearendingmarch20
19toyearendingmarch2020fromthecrimesurveyforenglandwales
csew

26 https://journals.sagepub.com/doi/pdf/10.1177/0269881115622241

27 https://pubmed.ncbi.nlm.nih.gov/33108459

28 https://pubmed.ncbi.nlm.nih.gov/33541760

29 https://www.aapcc.org/track/laundry-detergent-packets

30 https://www.reuters.com/article/us-health-poisoning-detergent
-idUSKCN1T52ZR

31 https://pubmed.ncbi.nlm.nih.gov/33636088

32 https://pubmed.ncbi.nlm.nih.gov/29555335

33 https://www.ncbi.nlm.nih.gov/pmc/articles/PMC7385722

34 https://pubmed.ncbi.nlm.nih.gov/30916627

35 https://academic.oup.com/DocumentLibrary/humrep/PR_Papers/
dez002.pdf

36 https://www.liebertpub.com/doi/10.1089/can.2020.0065

37 https://www.vox.com/science-and-health/2018/11/20/18068894/
marijuana-pregnancy

38 https://jamanetwork.com/journals/jama/fullarticle/2736583

39 https://reproductive-health-journal.biomedcentral.com/articles/
10.1186/s12978-020-0880-9

40 https://elearning.rcog.org.uk/sites/default/files/Domestic%20abuse%20
and%20substance%20misuse/BJOG_Fergusson_2002_0.pdf

41 https://www.ncbi.nlm.nih.gov/pmc/articles/PMC5812006

42 https://www.medicalnewstoday.com/articles/327230

43 https://www.health.harvard.edu/staying-healthy/pot-smokers-can-
maybe-breathe-a-little-easier

CHAPTER FOURTEEN

1 https://www.nytimes.com/2016/07/13/nyregion/k2-synthetic-marijuana-overdose-in-brooklyn.html
2 https://www.talktofrank.com/drug/synthetic-cannabinoids
3 http://neptune-clinical-guidance.co.uk/wp-content/uploads/2016/07/Synthetic-Cannabinoid-Receptor-Agonists.pdf
4 http://www.dldocs.stir.ac.uk/documents/rdsolr0305.pdf
5 Daly, M. (2014). 'Down a stony road: The 2014 DrugScope Street Drug Survey.' Available at http://www.drugsandalcohol.ie/23282/1/DownAStonyRoadDrugTrendsSurvey2014.pdf [accessed 7 December 2015].
6 https://www.theguardian.com/society/2018/sep/01/rise-in-prison-officers-contraband-smuggling
7 https://www.independent.co.uk/news/uk/home-news/spice-legal-highs-prison-criminals-deliberately-get-arrested-sell-drugs-jail-smuggled-inside-inmates-addicted-mamba-cheshire-police-expert-detective-constable-jamie-thompson-a7955271.html
8 https://www.justiceinspectorates.gov.uk/hmiprisons/media/press-releases/2016/07/hm-inspectorate-of-prisons-annual-report-201516-prisons-unacceptably-violent-and-dangerous-warns-chief-inspector
9 https://assets.publishing.service.gov.uk/government/uploads/system/uploads/attachment_data/file/669541/9011-phe-nps-toolkit-update-final.pdf
10 https://www.independent.co.uk/news/uk/home-news/legal-highs-spice-mamba-nps-drug-deaths-in-prison-triple-alarming-increase-war-on-drugs-overdose-psychotic-episodes-suicides-novel-new-psychoactive-substances-fake-weed-cocaine-heroin-prison-ombudsman-nigel-a7325666.html
11 https://www.theguardian.com/commentisfree/2018/may/16/spice-epidemic-uk-prisons-nurses-psychoactive-drugs-jails
12 https://assets.publishing.service.gov.uk/government/uploads/system/uploads/attachment_data/file/119149/acmd-report-agonists.pdf
13 https://assets.publishing.service.gov.uk/government/uploads/system/uploads/attachment_data/file/119042/synthetic-cannabinoids-2012.pdf
14 https://assets.publishing.service.gov.uk/government/uploads/system/uploads/attachment_data/file/380161/CannabinoidsReport.pdf

15 https://www.thelancet.com/pdfs/journals/lancet/PIIS0140
6736(17)31461-7.pdf
16 https://www.thelancet.com/journals/lancet/article/PIIS0140-
6736(17)31461-7/fulltext
17 https://www.ajmc.com/view/cannabis-shown-to-
relieve-parkinson-disease-symptoms
18 https://www.gov.uk/government/news/new-crackdown-on-danger
ous-legal-highs-in-prison
19 https://data.justice.gov.uk/prisons/prison-reform/random-manda
tory-drug-testing
20 https://assets.publishing.service.gov.uk/government/uploads/system
/uploads/attachment_data/file/873344/hmpps-annual-digest-2018-
19-march-2020-update.pdf
21 https://www.huffingtonpost.co.uk/entry/spice-in-manchester-
psychoactive-substances-bill_uk_58efb7b7e4b0da2ff85f09af
22 https://pubmed.ncbi.nlm.nih.gov/26577065
23 https://www.gov.uk/government/organisations/advisory-council-on
-the-misuse-of-drugs
24 https://www.thelancet.com/pdfs/journals/lancet/
PIIS0140-6736(17)31461-7.pdf
25 http://www.drugscience.org.uk/blog/2017/1/24/thcv-an-abject-cop-
https://www.drugscience.org.uk/thcv-an-abject-cop-out-
from-the-acmd-another-research-opportunity-for-the-uk-lost

CHAPTER FIFTEEN

1 https://www.gov.uk/government/collections/drug-driving
2 https://www.rac.co.uk/drive/news/motoring-news/drug-driving-
convictions-quadruple
3 https://newfrontierdata.com/cannabis-insights/roadside-drug
-testing-proves-inconsistent-for-cannabis-consumers
4 https://www.ukdrugtesting.co.uk/pages/uk-drug-driving-tests
5 https://www.think.gov.uk/themes/drug-driving
6 https://www.gov.uk/government/collections/drug-driving
7 https://www.dtecinternational.com/news/drugwipe
8 https://analyticalsciencejournals.onlinelibrary.wiley.com/doi/
10.1002/dta.2687

9 https://www.nytimes.com/2018/11/02/world/canada/canada-pot-snowboarder-olympics.html
10 https://jamanetwork.com/journals/jama/article-abstract/358432
11 https://pubmed.ncbi.nlm.nih.gov/1267035
12 https://assets.publishing.service.gov.uk/government/uploads/system/uploads/attachment_data/file/167971/drug-driving-expert-panel-report.pdf
13 https://pubmed.ncbi.nlm.nih.gov/30468948
14 https://bmjopen.bmj.com/content/9/5/e027432
15 https://newfrontierdata.com/cannabis-insights/data-shows-fatal-traffic-accidents-do-not-increase-after-cannabis-legalisation
16 https://www.ncbi.nlm.nih.gov/pmc/articles/PMC2722956
17 https://www.gov.uk/government/collections/drug-driving
18 https://newfrontierdata.com/cannabis-insights/roadside-drug-testing-proves-inconsistent-for-cannabis-consumers
19 https://www.ncbi.nlm.nih.gov/pmc/articles/PMC5890870
20 https://pubmed.ncbi.nlm.nih.gov/28531767
21 https://jamanetwork.com/journals/jamanetworkopen/fullarticle/2716990
22 https://www.ncbi.nlm.nih.gov/pmc/articles/PMC5890870
23 https://jamanetwork.com/journals/jamanetworkopen/fullarticle/2716990
24 https://assets.publishing.service.gov.uk/government/uploads/system/uploads/attachment_data/file/609852/drug-driving-evaluation-report.pdf
25 https://pubmed.ncbi.nlm.nih.gov/33790818
26 Ibid.
27 Ibid.
28 Ibid.
29 https://blogs.bmj.com/bmj/2013/03/27/kim-wolff-drug-driving-limits
30 https://mstrust.org.uk/sites/default/files/DfT-New%20Drug%20Driving%20Rules-Sativex-A5.pdf
31 https://www.cancard.co.uk/faqs
32 https://www.medicalnewstoday.com/articles/324315.php

CHAPTER SIXTEEN

1 https://yougov.co.uk/topics/politics/survey-results/daily/2021/04/06
   /fcf4a/3?utm_source=twitter&utm_medium=daily_
   questions&utm_campaign=question_3
2 https://gizmodo.com/norway-to-decriminalize-personal-
   purchase-possession-1846309869
3 https://eu.lohud.com/story/opinion/2021/01/11/ny-can-lead-nation
   -equitable-marijuana-legalization/6624819002
4 https://ballotpedia.org/Colorado_Proposition_AA,_Taxes_
   on_the_Sale_of_Marijuana_(2013)
5 https://pubmed.ncbi.nlm.nih.gov/29459211
6 Ibid.
7 https://www.sciencedirect.com/science/article/pii/S0955395918300
   264?via%3Dihub#fig0010
8 https://www.politics.co.uk/comment/2021/04/09/why-cannabis-
   reform-is-a-vote-winner-in-london
9 https://volteface.me/tory-mp-crispin-blunt-reveals-reasons-
   advocating-legalisation-cannabis-uk
10 https://www.manchestereveningnews.co.uk/news/greater-
   manchester-news/i-think-cannabis-should-legalised-15453597
11 https://www.theguardian.com/world/2020/nov/09/new-zealands-
   rejection-of-legalising-cannabis-is-a-triumph-for-fear-mongering
12 https://www.vice.com/en/article/bvxgn5/britains-second-largest-
   police-force-to-stop-criminalising-drug-users
13 https://www.theguardian.com/uk/2008/oct/03/blair.conservatives
14 https://www.london.gov.uk/what-we-do/mayors-office-policing-
   and-crime-mopac/our-priorities
15 https://www.healtheuropa.eu/police-boss-says-uk-should-regulate-
   cannabis-and-allow-home-grows/96868
16 https://prohibitionpartners.com/2021/02/12/cannabis-consump
   tion-europe
17 https://calgaryherald.com/news/cannabis-use-among-canadian-
   teens-down-since-legalization-say-researchers
18 https://jamanetwork.com/journals/jamapediatrics/fullarticle/
   2737637
19 https://www.forbes.com/sites/irisdorbian/2020/12/03/new-data-
   reveals-no-link-between-increased-cannabis-use-in-teens-and-legal
   -markets/?sh=1266520e3110

20 https://jamanetwork.com/journals/jamapediatrics/article-abstract/
    2754816
21 https://www.bloomberg.com/news/articles/2021-02-21/teen-use-
    mental-health-spur-talk-of-pot-curbs-cannabis-weekly
22 https://www.bbc.co.uk/news/world-us-canada-48921265
23 https://www.sos.state.co.us/CCR/GenerateRulePdf.
    do?ruleVersionId=8439&fileName=1%20CCR%20212-3

ACKNOWLEDGEMENTS

Thanks to my wonderful writer – Brigid Moss – without whom this would not have happened. To the Drug Science team for all their work over the past decade exploring the scientific truth about cannabis, publishing definitive papers on effects and harms and setting up the unique medical cannabis treatment initiative Twenty21. My own research on cannabis has relied on the inputs of Dr Robin Tyacke, Dr Matt Wall and especially Prof Val Curran.

Thanks also to Dr Susan Weiss for her careful and critical reading of the first draft and to Prof Dave Finn, who has clarified many technical questions relating to the endocannabinoid system.

Cali Mackrill did a great job in producing the figures.

A special mention is needed to all the patients with disorders that have benefited from medical cannabis who have been prepared to share their experiences despite having to access it illegally for the best part of their lives. I hope this book will go some way to removing the stigma of medical cannabis and facilitate the optimal use of this exciting new form of medicine.

INDEX

INDEX

ABOUT THE AUTHOR

DAVID NUTT
MBBChir (Cambridge), DM (Oxford),
FRCP, FRCPsych, FBPhS
FMedSci, hon DLaws (Bath)

David Nutt is a psychiatrist and the Edmund J. Safra Professor of Neuropsychopharmacology in the Division of Brain Science, Dept of Medicine, Hammersmith Hospital, Imperial College London. His research area is psychopharmacology – the study of the effects of drugs on the brain, from the perspectives of both how drug treatments in psychiatry and neurology work, and why people use and become addicted to some drugs such as alcohol. To study the effects of drugs in the brain he uses state-of-the-art techniques such as brain imaging with PET and fMRI plus EEG and MEG. This research output has led to over 500 original research papers that puts him in the top 0.1% of researchers in the world. He has also published a similar number of reviews and book chapters, eight government reports on drugs and 37 books, including one for the general public, *Drugs: Without the Hot Air*, that won the Transmission Prize in 2014.

He has held many leadership positions in science and medicine including presidencies of the European Brain Council, the British Association of Psychopharmacology, the British Neuroscience Association and the European College of Neuropsychopharmacology. He is currently Founding Chair of DrugScience.org.uk, a charity that researches and tells the truth about all drugs, legal and illegal, free from political or other interference. He also currently holds visiting professorships at the Open University and University of Maastricht.

David broadcasts widely to the general public both on radio and television and has a podcast (see below). In 2010 *The Times Eureka* science magazine voted him one of the 100 most important figures in British science, and he was the only psychiatrist in the list. In 2013 he was awarded the John Maddox Prize from Nature/Sense about Science for standing up for science and in 2017 a Doctor of Laws hon causa from the University of Bath.

en.wikipedia.org/wiki/David_Nutt
www.sciencemag.org/content/343/6170/478.full
https://www.imperial.ac.uk/people/d.nutt
https://www.imperial.ac.uk/medicine/departments/
www.drugscience.org.uk/drug-science-podcast

If you enjoyed reading *Cannabis*, why not read Professor David Nutt's book *Drink? The New Science of Alcohol + Your Health?*

yellow
kite

books to help you live a good life